Animal Behavior Science Projects

Best Science Projects for Young Adults

Animal Behavior Science Projects, Nancy Woodard Cain
Plant Biology Science Projects, David R. Hershey

BEST SCIENCE PROJECTS FOR YOUNG ADULTS

Animal Behavior Science Projects

NANCY WOODARD CAIN

John Wiley & Sons, Inc.
New York · Chichester · Brisbane · Toronto · Singapore

This text is printed on acid-free paper.

Published by John Wiley & Sons, Inc.

The publisher and the author have made every reasonable effort to ensure that the experiments and activities in the book are safe when conducted as instructed but assume no responsibility for any damage caused or sustained while performing the experiments or activities in this book. Parents, guardians, and/or teachers should supervise young readers who undertake the experiments and activities in this book.

Library of Congress Cataloging-in-Publication Data:

Cain, Nancy Woodard.
Animal behavior science projects / Nancy Woodard Cain.
p. cm. — (Best science projects for young adults)
Includes bibliographical references and index.
ISBN 0-471-02636-0 (acid-free : pbk.)
1. Animal behavior—Experiments—Juvenile literature. 2. Zoology projects—Juvenile literature. [1. Animals—Habits and behavior—Experiments. 2. Experiments. 3. Science projects.] I. Title. II. Series.
QL751.5.C295 1995
591.51'078—dc20 94-21779

Printed in the United States of America
10 9 8 7 6 5 4 3 2 1

For Alex,
who loves watching squirrels
hide peanut butter sandwiches in the trees

Acknowledgments

A number of individuals helped in the preparation of this book. The author would like to thank her husband, Tom, for his excellent critiques, patience, and devotion to fatherhood throughout the project. Jeanne Walsh of Dover Free Library was a tremendous research resource, and the staff of Marlboro College's library was always there to help me when needed. Diane Ciemniewski worked endless hours preparing detailed illustrations. Kate Bradford, John Cook, and the rest of the staff at Wiley did a highly professional job with the manuscript.

Contents

How to Use This Book

Included in this book are 20 projects on observing animal behavior. They are projects you can do in your home, in a zoo, at a city park, on a farm, in a pet store, or any other place where animals are found. The projects are organized into four main research areas:

1. Talking Animals: Includes projects on how animals, such as fireflies, wolves, and fish, communicate with one another.
2. Mother and Child: Includes projects on how mother animals, such as horses, ducks, and humans, interact with their children.
3. Play and Social Behavior: Includes projects on how animals, such as chickens, dogs, and pigeons, relate to each other in day-to-day situations.
4. Eating and Sleeping: Includes projects on how animals, such as spiders, squirrels, and cats, find food and where they make their homes.

Look through the table of contents for a topic or an animal that interests you. You may find that you are interested in a particular project but cannot find the animal to observe. If that is the case, check the "Animal Cross-Reference" in the Appendix for other species with similar behaviors.

But before you go directly to your project, read the next three sections on "The Art of Observing Animals," "Animal Behavior and the Scientific Method," and "Ethics in Animal Behavior Research." The information in these sections forms the basis of any animal behavior research project.

The Art of Observing Animals

Did you ever wonder why a dog tucks its tail under as it approaches its master? Or why a cat twitches its tail when hunting prey? Do you know that only the male cricket sings? Or that male mallard ducks don't quack—they whistle? Did you know that you can distinguish a male human infant from a female human infant on the basis of its smile? These are but a few of the questions and surprising findings that await you in the study of animal behavior.

People see animals all the time. Few people actually observe them. Even fewer people take the time to understand their behavior. There is an art to observing animal behavior that is a little like being a detective. You see an animal doing something, and you wonder why. First you learn about the individual animal. Then you do a background check: Where does it live? What are its habits? Armed with this information, you begin your investigation.

WHAT IS IT?

It is hard to observe an animal unless you know what you are observing. At first, it is not easy to tell the difference between animals of the same **species**. But the longer you watch them, the easier it gets. Look for physical differences. Are some larger than others? Are they different colors? Do they have different markings? Chipmunks, for example, have tails as different as human faces. Some are short, some are long, some are thick, some are thin, some are tapered, some are flat, and some may be missing hair.

Look for differences in behavior. Are some animals doing different things than others? For example, are some always chasing others or being chased? Do some animals repeat the same behavior over and over? Individuals who work with animals all the time, such as zookeepers, get to know them individually. If you ask a zookeeper how to identify individual zebras in a herd, he or she will point out zebras with wider stripes, zebras with thicker manes, and zebras with longer legs (Figure I.1). But they probably also will tell you each animal's name and all about their individual personalities!

Look for differences between males and females. It may be easy to tell the males from the females if they are physically different, or if they have distinct markings or coloring. In many species, however, the only clue to their sex is in their behavior. For example, the only way to tell a male pigeon from a female pigeon is by observing their courtship behavior.

Look for differences in age. Do some animals appear to be younger than others? Are they smaller or differently colored? Are they nursing or being held by the mother? Often you will see differences in how young animals interact with their peers and older animals. For example, puppies may engage in rough and tumble play with each other but avoid social interactions with older males.

Figure I.1. Physical differences in a herd of zebras.

WHERE DOES IT LIVE?

Where do you look for animals? The best place is where they live. When you see an animal in a natural setting, you are most likely in its **home range**. An animal's home range is the area in which it is born, lives, eats, sleeps, and dies. A beaver, for example, spends its entire life within a couple hundred yards of its lodge. In contrast, a black bear may range over an area of up to 175 square miles (453 square km).

The size of an animal's home range depends on many factors. These include food and water sources, nesting sites, and the season of the year. Home ranges for different species of animals may also overlap (Figure I.2). In one area, you might find mice, moles, chipmunks, gray squirrels, raccoons, and deer, among other animals.

Try to understand the boundaries of an animal's home range. The boundaries often include natural barriers such as a stream, the border of a lawn, or the edge of a forest. Observe where the animal eats, where it goes for water, and where its nest is located.

Many studies of animal behavior are conducted in zoos. A well-designed zoo exhibit mimics an animal's home range on a smaller scale. It may include rocks, caves, and trees for nesting sites and hiding areas. But behavior you observe in a zoo is often different from that seen in the

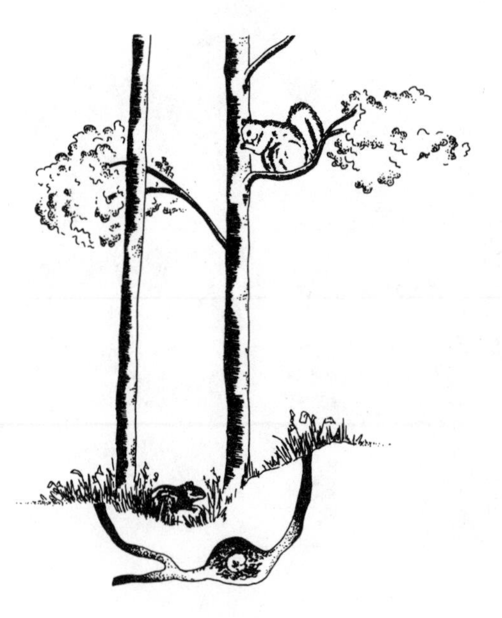

Figure I.2. Overlapping home ranges.

wild. This may be because the area is smaller or food sources are different. It also may be because there is little or no competition among animals.

WHAT DOES IT DO?

All animals, including humans, have certain habits. When is an animal active? When and where does it eat? How does it move about? The more you understand about these habits, the easier it is to observe behavior.

It is best to observe animals during their most active time of day. Otherwise, you may spend your time searching for an animal or watching it sleep. Some animals are active only at night. Others are active during the day. Animals also may have periods of peak activity. For example, some species of fireflies are active just before sunset and other species are active around dusk. For most of the year, gray squirrels are active just after sunrise and again during midafternoon. Yet in winter, they are active only around noontime.

Animals spend a lot of their time eating. For mammals, this is their most common activity. Animals often come out in the open in search of food, so they are easier to see. They also frequently return to the same feeding spots at the same time of day. Once you know an animal's feeding habits, you can plan your observation times around them.

Much like humans, many animals follow certain paths or routes to get from one place to the next. If you understand how an animal moves about its space, you do not have to search for it. All you need to do is check its route. Cottontail rabbits, for example, create narrow trails through shrubbery and tunnels through tall weeds. They stick close to these trails and dash to the nearest trail when in danger. Deer also create well-worn paths from feeding sites to hiding places. The same paths may be used for years.

When you have gathered background information on an animal, you are ready to begin a **systematic study**, or a planned observation, of its behavior. Where do you begin? The next chapter explains how to organize your study of animal behavior.

Animal Behavior and the Scientific Method

The deep blackness against the new fallen snow caught my eye. It paused, then streaked across the rise and was gone. I'd never seen anything like it. Was it a mountain lion or a bobcat? No, it wasn't catlike. Was it a fox, a wolf, or a coyote? No, it wasn't doglike. I hurried to get dressed to check its tracks. I compared its footprints to my pocket guide. Could it be a fisher? I'd never heard of one. I looked it up . . . a member of the weasel family . . . "secretive nature . . . rarely observed in the wild . . . even researchers have trouble locating fishers for firsthand observation." I was hooked. I read on and discovered that a fisher moves about its home range of five to eight square miles following a somewhat circular path. Good chance I'll see it again, I thought. And I did, several days later. What an animal! I looked for it every morning at sunrise and watched it all winter. It was a very special experience for me, for I had seen and learned about something most people never see in a lifetime.

This firsthand observation of the fisher illustrates much about the **scientific method** used in animal behavior. I did not conduct a classical experiment. I just watched the animal from my window. Yet, intuitively, I followed the scientific method. I *observed* an event, the appearance of an animal I did not recognize. I *questioned* my observation and drew upon my knowledge of other animals as I tried to classify it. When I did not have an answer, I looked for additional cues—I checked its tracks. Then I *researched* the animal and compared what I read with what I observed. I read what little we know about its environment and behavior. I *hypothesized* I would see it again at the same time of day and in the same place. I waited for it and *tested* my hypothesis. I *recorded* what I had seen and when. I *discovered* a new animal that winter. And in the process, I had a rare glimpse into the private life of an elusive animal.

Observe, question, research, hypothesize, test, record, and discover. This is what the scientific process is all about. You follow the same approach for **field studies** that you do in the laboratory. Observing animal behavior in the field is a little more demanding. You must plan for the unexpected. You need to organize the mountains of data you collect. And you must always fight to remain objective, to simply record what you see. In this section, you will discover how to observe animal behavior using the scientific method.

OBSERVING

How do you begin a study of animal behavior? You make some general observations, like those outlined in the chapter "The Art of Observing Animals." General observations are important for two reasons: They help you focus your study, and they provide you with essential background information.

As observers, we tend to see what we want to see. Consider Figure I.3. What does it look like? Most people will answer that they see a brick wall, a tile floor, a rug, a block pattern, or some other similar object. If, however, you are told to count the number of triangles in the design, you see something entirely different. You see just a pattern of triangles and rectangles.

This example illustrates the difference between general observations and focused observations. Observing animal behavior involves finding the right balance between the two. Start out with general observations of an animal or its behavior. This will allow you to consider all possibilities

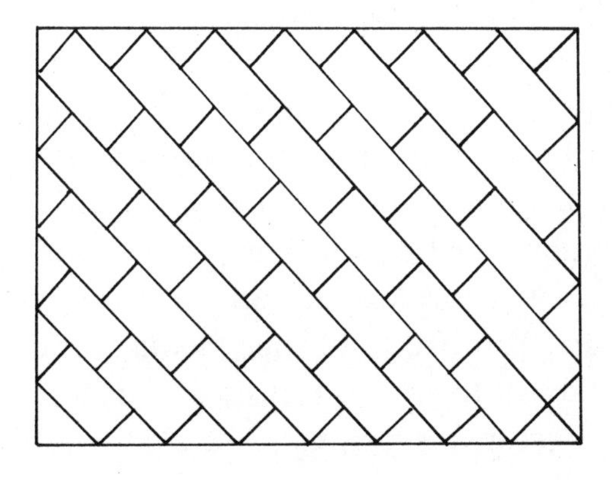

Figure I.3. We see what we are looking for.

for study. As you try to make sense of what you see, focus on the behaviors you do not understand. Focus on those behaviors that intrigue you. These are the behaviors that will form the basis of your study.

QUESTIONING

As you observe behavior, constantly question what you see. Write these questions down for later use. Do not make any assumptions, and do not take anything for granted. Above all, guard against **anthropomorphism,** or the tendency to attribute human characteristics to nonhuman animals. It is easy to fall into that trap, such as asking, "Did the squirrel go into its nest because it wanted to escape from the cold?"

Try to keep your questions open and objective. Ask questions for which the answer is observable. Do not ask, "Did the squirrel go into its nest because it was hungry?" Ask, "When does a squirrel go to its nest?" or "Does the squirrel store food in its nest?" A few observations will answer your questions. Watch the squirrel eating. Does it eat as it collects food? What does it do with food it does not eat? Does it carry the food somewhere else? Does it bury it? Do you ever see the squirrel carrying food to its nest? Unless you climb a tree and check inside a nest, you will never know without a doubt whether the squirrel stores food there. However, your observations will tell you the *likelihood* that a squirrel stores food in its nest. (By the way, squirrels bury foods such as acorns and nuts. They only sleep in their nests at nighttime.)

RESEARCHING

The more you know about an animal, the more you can learn by observing it. For many animals, there is a limit to what you can easily observe. You cannot follow an animal like a black bear around its home range of a 100 or so square miles (259 square km)—nor would it let you! If you observe a black bear in a zoo, you see only a sliver of its life in the wild. So what do you do? You use published books, articles, and papers to research the behavior of the animal further. This information is used in planning and organizing your study. You also use this information to interpret your results.

People who study the behavior of animals include **wildlife biologists, psychologists, ethologists, zoologists,** veterinarians, zookeepers, hunters, trappers, farmers, and animal trainers. These individuals write books and articles on animal behavior topics of interest to them. A good place to start your research is with a general book on animal identification or animal behavior. Books such as *Animal Tracking and Behavior,* by Donald and Lillian Stokes, or *Eastern Forests,* by Ann and Myron Sutton, are examples of excellent overviews. Some books are written exclusively on one type of animal, like *Horsewatching,* by Desmond Morris, or *Chickens, Chickens, Chickens,* by Peter A. Limburg.

People studying animal behavior also publish general articles on animals. Articles appear in magazines such as *National Geographic, Scientific American, Wildlife Conservation,* and *Field and Stream.* These articles are usually of popular interest and provide excellent background information on an animal.

More specific information on an animal or on some aspect of its behavior can be found in **journal articles**. Journal articles describe detailed experiments or field studies conducted by professionals. Examples of professional animal behavior journals include *Behaviour, Animal Behaviour,* and the *Journal of Comparative Psychology.*

Ask your librarian to assist you in researching the behavior of a particular animal. Your school library will probably have a section of books on domestic and wild animals. Use the public library to search for magazine articles. If the library does not have the magazine you need, ask if the librarian can order it for you. Most libraries have an arrangement with a larger or regional library and can get an article for you in a few days.

FORMING A HYPOTHESIS

Once you have selected a topic for study and have done some background research, you are ready to outline your project. Initially, your objective may be to simply observe behavior. Suppose you observe the courtship behavior of pigeons. While observing, you undoubtedly have questions. And you find yourself naturally making predictions. For example, you may observe a pigeon repeatedly bowing its head and dragging its tail. What does it mean? You watch a little longer. Maybe it is a behavior performed by the male to initiate courtship? Your questioning has led to a prediction, or a **hypothesis.** A hypothesis is an educated guess, based on your preliminary observations.

All experiments and many observational studies begin with a hypothesis. The hypothesis is a statement of belief; it must be testable or observable. For example, in Chapter 11, it is hypothesized that neon tetra fish school using a combination of chemical and visual communication.

TESTING

Is your hypothesis correct? Hypotheses are tested in two fundamental ways. You can use the scientific method to create the behavioral events to be studied, or you can use **naturalistic observations** to study the behavioral events as they occur in a natural setting. In either case, the design of your project should test your hypothesis.

The Scientific Method

The scientific method involves testing a prediction by manipulating certain factors called **variables** while holding all other factors (called **controls**) constant. For example, in Chapter 4, a male Siamese fighting fish is presented to three different types of opponents to see which one elicits the most aggression. In this experiment, the type of opponent is the variable. However, because the same fish is exposed to three different opponents, it may learn to be more or less aggressive through repeated aggressive encounters. The effects of repeated experience are controlled by presenting each opponent three different times, varying the order each time they are presented.

Good experiments are both well controlled and **replicable.** Any researcher should be able to re-create your experiment and repeat your results by following your procedure. A result that cannot be repeated is meaningless. Thus, it is essential that an experiment be well documented. This includes a description of the materials and precise procedures required to conduct the experiment. No details can be assumed.

Naturalistic Observations

Naturalistic observations are used to record behavior as it occurs in a natural setting. Sometimes observations are made to test a hypothesis.

Other times, observations are made to gain further understanding of an animal or a behavior. In its simplest form, naturalistic observation involves recording everything that happens. There are a lot of pitfalls to recording everything you see. Things you do not see might be significant. Other people doing a similar observation may not see what you see.

There are techniques you can use with naturalistic observations to make your observations more reliable. For example, suppose you want to know how much of the day an animal spends sleeping. You could observe the animal continuously for many days, under many conditions, or you could observe a **sample** of its behavior and estimate the amount of time it spends sleeping. This technique is used in Chapter 6 to answer the question, "How close does a mother horse stay to her foal during its first eight weeks of life?"

The Scientific Method versus Naturalistic Observations

When studying animal behavior, there are distinct advantages and disadvantages to the scientific method versus naturalistic observation methods. There is no question that the scientific method allows far greater control. However, many questions in animal behavior are difficult or impossible to study in the laboratory. For example, Chapter 1 investigates how different species of fireflies communicate with flash signals in two different **habitats** (environments) and at different times of the evening. This would be a very difficult study to replicate in the laboratory. And, most likely, it would lead to different conclusions. In general, studies that relate an animal's behavior to its habitat must be done in naturalistic settings.

RECORDING THE DATA

If I could give new students of animal behavior one piece of advice, it would be to construct a well-organized form for collecting **data**. As you design your experiment, think about what you need to observe to test your hypothesis. Spend some time planning your **data sheet**, the form you use to record your observations. Design a form that requires a minimum amount of writing. For example, if you are observing hunting behavior of domestic cats, you could write down every behavior you see. Or, based on your preliminary observations and research, you could construct a data sheet with columns for each of the behaviors you might see. Then, you would just place an X in the appropriate column as you observed each behavior.

For example, consider the data you would collect in Chapter 20 on bumblebee searching strategies. Testing the hypothesis in this experiment involves making two observations: What type of flower does a bumblebee land on? Does it land on the edge or center of the flower? You could simply make a number of entries down the page, indicating the type of flower the bee landed on and where it landed.

Alternatively, you could design a data sheet like the one that follows after determining the various flower possibilities. Such a form makes data collection much more efficient. You do not miss significant behaviors while recording others. Organized data sheets also make analysis of the data much easier and more efficient. To summarize the following data, all you need do is sum the entries in each column and calculate the percentages.

<table>
<tr><td colspan="4">Sample Data Sheet
Bumble #1</td></tr>
<tr><td colspan="4">Date: ____________
Begin collection trip: ____________
End collection trip: ____________</td></tr>
<tr><td>Wild Rose</td><td>Daisy</td><td>Buttercup</td><td>Flowering Quince</td></tr>
<tr><td>EEEEE
EEEEE</td><td>EC</td><td>C</td><td></td></tr>
<tr><td colspan="4">Notes: An "E" is used to indicate that the bee landed on the edge of the flower; a "C" is used to indicate that the bee landed on the center of the flower. It is easier to tally your data if you record instances in blocks of five, as was done in the column for wild roses.</td></tr>
</table>

DISCOVERING

One of the most exciting parts of the scientific method is discovering something you did not know before. It may just be a new species of bird you have never seen before. It may be a behavior you have seen but never understood. Or it may be a behavior you have never noticed. In the end, you have recorded a lot of descriptions and results. What do they all mean?

Summarizing your data is the first step in understanding its meaning. As shown above, a well-organized data sheet will simplify the process. Think about all of the questions raised by your observations. Then, ask yourself how you can analyze your data to answer those questions. Go back to your library research and see what other people have learned about the same or similar animals. Based on your observations and the observations of others, how should you interpret your results? Do your results suggest the need for further investigations?

Ethics in Animal Behavior Research

As you observe behavior, either human or nonhuman, you have a responsibility to be aware of ethical considerations. If you work with animals in your home, treat them humanely and with respect. If you observe animals in the wild, take care not to disturb their habitat or alter it in any way. If you observe humans, make sure that you thoroughly explain to them the purpose of your study (to the extent you can without affecting their responses). After the study is over, **debrief** all of your human subjects by explaining the purpose of your study and why you used certain observational methods.

Many science fairs have formal guidelines or rules. Be sure you are informed about any policies on experimentation with animals. Science fairs usually welcome projects like the ones in this book that do not harm animals.

PART I

Talking Animals: Projects on Animal Communication

Animals communicate in many ways. Just like humans, each species of animal has its own language. Often we do not understand the meaning of these communications because they are so different from our own. For example, honeybees communicate using odor, sound, contact, and vision. They also respond to gravity and vibrations. It is hard to imagine trying to say, "Pass the butter please," using odor or vibrations to communicate!

Why have nonhuman animals developed such complex forms of communication? Varied forms of communication are necessary for survival. Some animals need to communicate in darkness. Others need to communicate without being seen by their enemies. Many animals need forms of communication that bypass obstacles or can be sensed in an underwater environment. Still others must be able to communicate over long distances.

What are some forms of communication used by animals? Animals can perform flashy **aggressive displays** or **courtship displays.** Many have a wide range of facial expressions and body postures. They can have an array of vocal communications, including **ultrasonic calls** that cannot be heard by the human ear. Many animals also use **chemical signals,** such as odor, to communicate. Physical contact is another common form of communication. And some animals even communicate using vibrations and **electrical signals.**

Animals, including humans, often use more than one sense in a single communication. For example, we might communicate alarm by saying "I'm afraid" and by a look of fear on our face. In the same way, a white-tailed deer communicates alarm by using three different senses.

First, it stamps its hooves, which sends vibrations a great distance through the ground. Then it snorts. Finally, it flags nearby deer with the white underside of its tail. By using three ways to communicate alarm, the deer ensures that its communication is "heard" by others near and far.

As you observe animals, try to understand their language. How do they react to your presence? How do they "talk" to each other? What sounds do they make? What visual signals do they use? Do they come in physical contact with each other? Do they sniff each other or mark objects with their scent? Observe closely. Try to tell what draws one animal to another.

CHAPTER 1

Firefly Blinking Patterns

Fireflies use a complex form of flash communication to attract mates. The glow, which comes from light-producing cells on the firefly's underside, is turned on and off at will. The flashing of fireflies seems random at first. But when you observe more closely, you can detect a pattern in the flashes.

Males signal to females while flying (Figure 1.1). Females respond to these signals from on or near the ground. When a female responds, the male continues to signal and flies closer. The male and female signal back and forth until the male locates the female. Then he mates with her.

There are a number of different species of fireflies. With fireflies flashing everywhere, how do fireflies find one of their own kind to mate with? Each species of firefly speaks its own language and has a unique flash pattern. The flash signals may vary in a number of ways, such as the color and duration of the flashes, the time between signals, and the number of flashes in a signal.

Figure 1.1. Male firefly flashing to a female near the ground.

Different species of fireflies also tend to live and mate in different settings. For example, one kind might live in meadows, another in wooded areas. Also, different species of fireflies are not all active at the same time of night. Some species start signaling around sunset, and others may not be active until after dusk.

Fireflies make themselves easy to spot. You can see them in fields, meadows, lawns, along hedges, in the woods, at the edge of the woods, and even on some city sidewalks.

Purpose

To observe how different species of fireflies communicate with flash signals.

Observational Setting

Choose two different settings in which to observe fireflies. Your results will be most interesting if these settings are near one another.

Materials

- ☐ notebook
- ☐ pencil
- ☐ stopwatch, or watch with second hand

Procedure

1. *Night 1.* Observe firefly behavior and try to understand how they use signals to communicate. Start your observation about 30 minutes before sunset in either of the locations you chose. Locate a male. Male fireflies signal while flying. Try to follow it and observe how it locates a female. Observe how the male signals to the female, as well as the female's response. Try to follow at least five different males.
2. *Night 2.* Start your observation about 30 minutes before sunset in the first of the settings you chose. Watch for signaling males and record the following:
   - the duration of each flash
   - the number of flashes in a signal

Try to observe as many fireflies as you reliably can. Continue observing until about 30 minutes after dusk. You might want to construct a data sheet like the following one to record your observations.

Setting: ____________			
Before Sunset		After Sunset	
No. of Flashes	Duration	No. of Flashes	Duration

3. *Night 3.* Prepare another data sheet and repeat your observations of the previous evening in the second setting you chose.

Questions

1. As you observed the male, did it fly around before focusing on a single female? Did it continue to flash as it flew closer?

2. Did you see more than one type of flashing pattern? You observed fireflies in two different settings and at two different times of the evening (before and after sunset). Chances are that you observed more than one type of flashing pattern.

3. Compare the results you obtained in the two settings. Was there a difference in the number of flashes you observed? Was there a difference in the duration of the flashes?

4. Compare the results you obtained before sunset and after sunset. Were there differences in the flash patterns you observed during the two time periods?

5. Did you notice any difference in the color of the flashes in either setting? During different times of the evening? Different colors may be a further clue that you have observed several types of fireflies. Some scientists have correlated the color of the flash with the time of the evening. They have found that fireflies with mostly yellow hues are seen at twilight. Later on in the evening, fireflies with darker, greenish hues are more likely to be seen. Why do you suppose yellow hues are seen at twilight, and more greenish hues are seen as it gets darker?

6. Each species of firefly has developed its own signaling system. If you saw differences in the number, duration, or color of flashes, you probably observed different types of fireflies. How many species of fireflies do you think you saw? Were any of them active at the same time and in the same setting?

CHAPTER 2

Eye Contact in Primates

The eyes are our most important body part when it comes to **nonverbal communication.** The emotional content of our communication is transmitted in large part by the nuances of the eyes and the use of the facial muscles around them.

Nonhuman primates, particularly species such as chimpanzees, gorillas, and orangutans, use the eyes to communicate in much the same way as humans. And, like humans, monkeys and apes spend little time in direct eye-to-eye contact.

The purpose of direct eye contact can seem contradictory. Direct stares signal both active aggression and active fear: highly aggressive animals stare directly at their opponent, whereas highly fearful animals do not take their eyes off their opponent. In a similar manner, looking away is used to signal **dominant** as well as **subordinate** behavior. Dominant animals look away from or ignore less important animals. Submissive animals avoid making eye contact.

How, then, do you tell what the eyes are saying? It is the accompanying facial expressions and body language that signal the emotion involved. It is also important to look at the context in which the behavior occurs. For example, a direct stare by the most dominant monkey in a troop can result in a fight. In contrast, a direct stare by the bottom-ranking individual will most likely be ignored by the other monkeys.

This experiment will focus on one type of eye contact: the threat display, which uses direct eye contact to convey **social status.** The threat display usually includes some of the following facial expressions and behavioral elements (Figure 2.1):

Mouth. Slightly to fully open. Upper lip rounded over teeth, which are usually not visible unless threat display escalates. Corners of mouth brought forward.

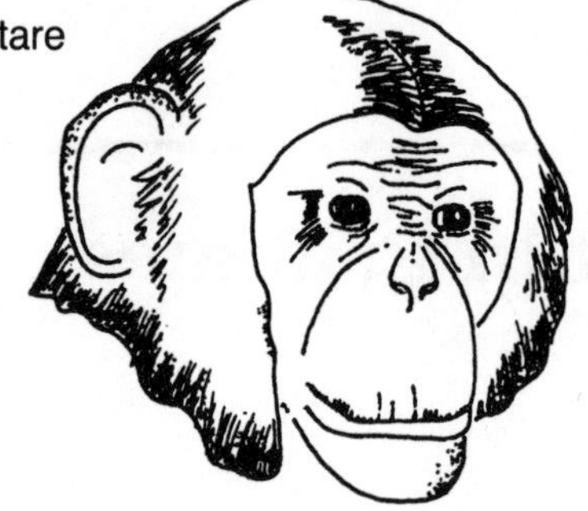

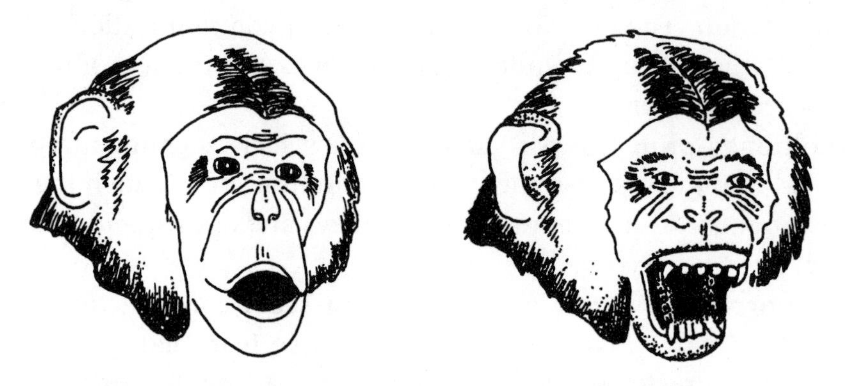

Figure 2.1. Threat-related facial expressions of chimpanzees.

Eyes. Fixed direct gaze. Vertical movement of eyebrows. Possibly rapid blinking.

Ears. Ears flattened against side of head. May be forward at onset of threat.

Nose. Nostrils may be flared.

Body postures. Slapping, lunging, or jerking.

The recipient of a threat display may respond in a number of ways, depending on the recipient's social status and the context. For example, the recipient may physically withdraw or simply look away. It also may respond with a submissive facial expression. It may stare directly at its opponent with a fearlike grimace in which its teeth are horizontally bared. The recipient may also crouch, scream, stare at another animal, or, rarely, threaten back.

Purpose

To observe how gorillas, chimpanzees, or orangutans use and respond to a threat display.

Observational Setting

This project must be performed in a zoo setting. An older zoo, in which you can closely approach the animals in cages, will produce the best results. You will probably get slightly different results if you perform this project in newer zoos where the animals live in more natural habitats. This is because the animals are separated from visitors by embankments or water and are physically more distant from the observer. It is best if you can conduct this project over several days.

Materials

- ☐ notebook
- ☐ pencil
- ☐ stopwatch, or watch with second hand

Procedure

1. *Observing naturally occurring eye contact behavior.* Find a good place from which you can observe closely the interactions of two or more gorillas, chimpanzees, or orangutans. Watch their eye contact behavior closely for 2 hours. Record the following data associated with their eye contact:
   - the initiator
   - the recipient
   - the duration
   - facial expressions accompanying the eye contact
   - body postures accompanying the eye contact
   - response of the recipient
2. *Observing responses to your eye contact and threat displays.* Find a good place where you can easily approach, interact with, and observe a gorilla, chimpanzee, or orangutan. Ideally, you should be able to stand within 6 feet (2 m) of the animal. Plan to make your observa-

tions over a one- or two-day period, waiting approximately 30 minutes between observations. Move slowly up to the cage and select an individual animal for observation. Often, animals will approach the front of the cage, as if seeking stimulation, when visitors appear. You will be making the following observations:

- how the animal responds to a "blank" stare from you.
- how the animal responds to a "threat" stare from you.

The stares will vary in duration from very short to long, as outlined in the following table. Thus, you will be making six different observations.

Blank Stare	Threat Stare
2 to 3 seconds	2 to 3 seconds
10 seconds	10 seconds
30 seconds	30 seconds

a. *Blank stares.* Stand erect within 4 to 6 feet (1 to 2 m) of the animal. Try to relax your facial muscles as much as possible and stare blankly at the animal for 2 to 3 seconds. Then look away. Move back from the animal to record your observations. Wait approximately 30 minutes. Repeat the observation, staring at the animal for 10 seconds before looking away. Wait an additional 30 minutes. Repeat the observation again, staring at the animal for 30 seconds before looking away. Record the animal's response to each stare.

b. *Threat stares.* Stand within 4 to 6 feet (1 to 2 m) of the animal (Figure 2.2). Assume a threatening stance and facial expression as follows: Bend your knees slightly and lean toward the animal. Stare directly at the animal's eyes, furrow your eyebrows (with a vertical motion, as if angry), open your mouth, and stretch your lips tightly without showing your teeth. In short, make a mean, threatening face. Follow the same procedure outlined for blank stares in step 2a, except use your threat display face.

Figure 2.2. Human giving a threat stare to a chimpanzee.

Questions

Naturally Occurring Eye Contact Behavior

1. How frequently did you observe eye contact between animals? How long did the eye contact last?
2. Assuming you observed some eye contact, can you describe the context in which it occurred? Did accompanying facial expressions help you to understand the context or the purpose?
3. What was the typical response of an animal to a nonthreatening stare?
4. What was the typical response to a threatening stare?

5. Do you think you would observe the same behavior if the zoo exhibit were larger or more densely planted?
6. Do you think that animals show a greater variety of eye contact and facial expressions in the zoo or in the wild?

Blank Stare Eye Contact

1. Do you think the animal perceived your blank stare as a threat?
2. Did the longer duration blank stare threaten the animal any more?
3. Did the animal turn away or stare back at you during each of the observations?
4. Did the animal shift its body posture or change its behavior in response to your stare? Did it vocalize? Did it move closer or farther away?
5. Even though you made eye contact with the animal for 30 seconds, give several reasons why it might not perceive the stare as a threat.
6. Did you find it awkward or uncomfortable to sustain eye contact with the animal for long durations?

Threatening Eye Contact

1. Did the animal respond differently to the threatening eye contact? Were the longer durations perceived to be more threatening?
2. Did the animal sustain eye contact or look away?
3. Did the animal threaten you back?
4. Did the animal shift its body posture or change its behavior in response to your stare? Did it vocalize? Did it move closer or farther away?
5. Did you feel threatened by its response even though you were protected by a barrier?
6. Was it difficult or awkward for you to sustain the threatening eye contact for long durations?
7. Are you surprised that a nonhuman primate can "read" human facial expressions and postures?

CHAPTER 3

Facial and Tail Expressions of Wolves

Wolves have a highly developed **social structure.** They mate for life and live in packs that include other family members and relatives. Wolves travel, hunt, and rest together as a pack. They feed on large mammals such as caribou, moose, and deer—animals too large to be brought down by a single wolf. Thus, their survival depends on cooperation. The pack does not hold together automatically. Rather, it is maintained by a tight social structure in which every wolf knows its place. Behind their social order is a highly developed system of communication.

The social behavior of wolves is based on a **dominance hierarchy** system of social organization. This system is best explained as a social ladder, in which each member occupies a certain position above or below another member. Animal A is dominant over animal B, who is dominant over animal C, and so on. The highest ranking or most dominant individual is called the **alpha animal.** The second highest ranking individual is called the **beta animal.**

A wolf pack actually has two separate dominance orders: a male order and a female order. The alpha male and the alpha female are usually the original mated pair. The remaining pack members are mature, subordinate animals; outcasts (males that stay near the fringes of the pack's social center); and juveniles. These pack members typically represent several generations of the mated pair's offspring. Only the alpha female produces offspring from year to year. Even though the subordinate females ovulate and court males, they do not actually mate.

The ability of pack members to communicate is essential to the maintenance of their social organization. Wolves communicate primarily through vision, smell, and hearing. The majority of their communication is visual, and the wolf's most important visual expression center is the head. The combination of the coloring of the face and the facial muscles controlling its eyes, ears, nose, and mouth creates a wide range of facial expressions. So expressive are wolves' faces that the highest ranking wolf can be identified by its facial expressions alone.

The facial expression of a dominant, threatening wolf is characterized by bared teeth, an open mouth with the corners pulled forward, a wrinkled and puffed-up forehead, and pointed, erect ears (Figure 3.1A). A high-ranking animal might also assert its dominance in a nonthreatening way by simply staring intently at the subordinate animal. The facial expression of a subordinate wolf is characterized by a closed mouth with the corners of the mouth pulled way back, a smooth forehead, slitlike eyes, and ears drawn back and held close to its head (Figure 3.1B).

Not only can you "read" a wolf's facial expression, but you can also read its tail expression. Several aspects of a wolf's tail, including position and movement, are important indicators of its mood and status. Using its tail, the wolf can affirm its social position or communicate its feelings from a distance. The position of the wolf's tail is a particularly good indicator of its social status (Figure 3.2). A dominant wolf carries its tail up high. A very subordinate wolf carries its tail between its legs or curved forward along its legs. A wolf of medium rank carries its tail somewhere in between.

Tail movements are a good general indication of the wolf's disposition. Wide-circled, loose tail wagging indicates friendliness. In very subordinate wolves, this typically takes the form of rump wagging. Tight-circled, quick tail wagging is associated with aggressiveness. A trembling tail indicates uncertainty, as when two high-ranking males meet.

Purpose

To determine the social organization of a pack of wolves using facial and tail expressions.

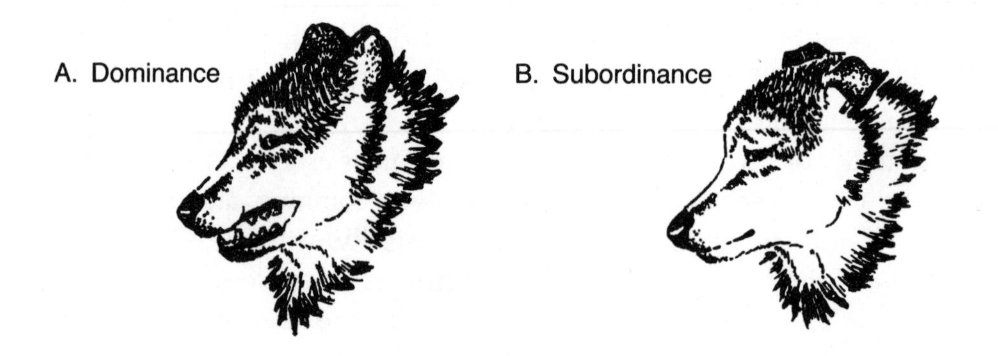

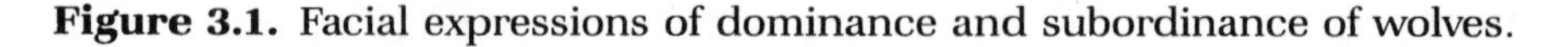
Figure 3.1. Facial expressions of dominance and subordinance of wolves.

Figure 3.2. Wolf tail positions signaling dominance and subordinance.

Observational Setting

This project must be done in a zoo setting. Only one pack of wolves will be in a particular exhibit. It is best done during colder weather, when you will see more activity among the wolves. Wolves typically mate between February and March. If you observe them during that time, you are likely to see a lot of activity surrounding mating. You probably also will see a greater number of social interactions around feeding time. Ask the zookeeper when the wolves are fed. Wolf cubs are born in late spring and early summer. Once they come out of their den, it is interesting to watch the young. It is also interesting to watch the rest of the pack's reaction to them.

Materials

☐ notebook

☐ pencil

Procedure

1. *Recognizing individual wolves.* Spend some time just watching the wolves in their exhibit. This will give you an idea of when they are active. It also will give you a chance to learn how to recognize each one individually. This may take a lot of patience and practice. Look for marks that will help you identify each animal. These might include the following:
   - size of the wolf
   - shape of its head or snout
   - differences in coloring
   - unique markings, particularly on its face and legs
   - shape of its ears
   - observable scars
   - quality of its coat

   If you have a difficult time recognizing individual wolves, ask the zookeeper how he or she tells the wolves apart. You will be amazed how easily they can identify their animals. If you also have a difficult time sexing the wolves, ask the zookeeper to provide this information as well.

2. *Mapping the dominance hierarchy.* Once you can identify each animal, you are ready to begin mapping out the dominance hierarchy. Using a data sheet like the one on page 31, record each instance of social interaction between wolves by using characteristics of their facial expression and tail position to identify the dominant and subordinate animal.

	Dominant	Subordinate
Face	Bared teeth	Closed mouth
	Open mouth with corners pulled forward	Corners of mouth pulled way back
	Wrinkled forehead	Smooth forehead
	Pointed ears	Slitlike eyes
		Ears drawn back
Tail	Carried high	Between legs

For example, if wolf A comes close to wolf B, and wolf B tucks its tail tighter between its legs, then you can assume that wolf A is dominant over wolf B. In the data sheet, the dominant animal is indicated on the vertical axis, or **y-axis,** and the subordinate animal is indicated along the horizontal axis, or **x-axis.** Thus, you would record a mark in the cell AB to indicate that A is dominant over B. Continue to record each social interaction this way until you have collected enough data so that a pattern emerges. Remember that there are two hierarchies, one male and one female, so you will need a different data sheet for each sex.

Male Wolf Interaction						
Date: ________ **Time:** ________						
		x—Subordinate Wolf				
		A	**B**	**C**	**D**	**E**
y—Dominant Wolf	**A**		X			
	B					
	C					
	D					
	E					

3. *Checking your data.* Once you have collected all of your data, ask to meet with the zookeeper to discuss your results. Ask the zookeeper to give you the ages and relationships of the wolves to each other. Check the reliability of your results by asking him or her to describe the pack's dominance hierarchies.

Questions

The Dominance Hierarchy

1. Did you correctly identify the alpha male and alpha female? Were these the two wolves with the greatest number of social interactions?

2. Did you ever observe any challenges to the alpha male and alpha female, or were they always greeted with submission? It would not be unlikely for you to see challenges, particularly if you observed the wolves during mating season.
3. Did the alpha male or alpha female seem to be the overall dominant member of the pack?

Facial and Tail Expressions

1. Did you find facial expressions or tail expressions to be better predictors of dominance and subordination?
2. Were tail expressions easier to detect than facial expressions?
3. Was dominant behavior easier to detect than subordinate behavior?
4. Why do you think that wolves evolved two different forms of communication—facial expressions and tail expressions—to signal the same information?

Other Forms of Expression

1. Did you observe other behaviors that were characteristic of either dominant or subordinate individuals? Try to describe them.
2. Did these behaviors accompany facial and tail expressions?

The Zoo Environment

1. Does the zoo setting or characteristics of the exhibit affect social organization? Is the alpha female still the only wolf that mates?
2. The zoo environment is much different from the wolves' natural habitat. It is smaller and has built-in features for shelter, and wolves are fed routinely, with no need to hunt. Do you think that any of these factors reduce their need to cooperate?

CHAPTER 4

Aggressive Behavior of Siamese Fighting Fish

Physical aggression and injury are rare occurrences in disputes over **territorial boundaries** between fish. Rather, conflicts are typically settled through aggressive displays and **ritualized fighting.** Aggressive displays are observed in many animal species. In fish, they usually involve a sudden increase in apparent size as the fish assume fighting postures. If one opponent is not intimidated, aggressive displays frequently escalate into ritualized fighting. Fish circle each other and beat the water forcefully with their tails, attempting to throw their opponent off balance.

One of the most well-studied aggressive displays is that of the male Siamese fighting fish (*Betta splendens*). Siamese fighting fish are members of the family Anabantidae, a group of small tropical fish found in Southeast Asia and Africa. These fish have a special breathing apparatus connected to their gill chambers that allows them to extract oxygen from the air. As a result, they can live in shallow, oxygen-deficient water that would be lethal for other fish.

Most of us know Siamese fighting fish as the beautifully colored, fresh water aquaria fish with long, veil-like fins. These are the domestic variety that have been carefully bred for their coloration and magnificent fins. Despite careful cultivation, they have retained much of their aggressive instinct.

The aggressive display of the male Siamese fighting fish looks something like the display of a peacock (Figure 4.1A). The fish erects its gill covers and extends its fins. To an opponent, it may appear twice its actual size.

Displaying fish alternate between facing each other and turning broadside to their opponent (Figure 4.1B). Facing is always accompanied

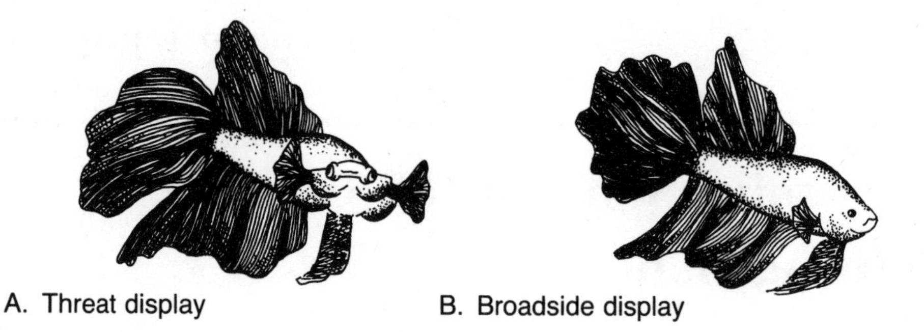

Figure 4.1. Threat display (A) and broadside display (B) of male Siamese fighting fish.

by gill cover erection. Turning broadside is accompanied by flickering of the pelvic fins and tail beating. The "duel" at its simplest level appears to involve tail beating in response to the erection of the rival's gill covers. The frequency of tail beating increases as the opponent's gill covers remain erect. Only when tail beating appears to reach its peak does the opponent lower its gill covers. The match gradually escalates until one fish gives up and stops tail beating.

Defeated male Siamese fighting fish attempt to flee and rapidly show a change in posture and coloration. The submissive fish assumes a tail-down posture. Its body and fins shift in color from a bright red, blue, or purple hue to a faded color. Prominent black horizontal bands appear on its body. Unless confined to a small area, the winner will often stop pursuing its submissive opponent.

As is the case with many species, male Siamese fighting fish display to females during courtship as well as to rival males. The typical courtship sequence begins very aggressively. The male displays to the female and chases it relentlessly. At the same time, the male builds a bubble nest for the eggs. Between chases, the male swims to the surface and blows bubbles, one at a time. These bubbles contain a small amount of mucus that makes them quite durable. An average bubble nest, which contains 400 to 500 bubbles, is several inches in diameter and one-half inch in depth. Once the female shows submissive behavior, the male continues to nip and butt her until she is ready to lay her eggs.

Purpose

To observe how a male Siamese fighting fish displays to different kinds of opponents.

Observational Setting

This project can be conducted in your home; however, you must be willing to take care of the fish once the project is done. You also may try to make arrangements to return the fish to the pet store. If you choose to keep them, read the "How to Care for Your Siamese Fighting Fish" section at the end of this chapter.

Materials

- ☐ 2 adult male Siamese fighting fish
- ☐ 1 adult female Siamese fighting fish
- ☐ 3 flat-sided, 1-quart (.95-l) glass fish bowls
- ☐ two 4-by-6-inch (10-by-15-cm) pieces of cardboard
- ☐ a mirror, approximately 5 inches (13 cm) square
- ☐ freeze-dried brine shrimp fish food
- ☐ 2 clean 1-gallon (4-l) plastic milk jugs
- ☐ notebook
- ☐ pencil
- ☐ stopwatch, or watch with second hand

Procedure

1. *Setting up the fish bowls.* Like other aquaria fish, Siamese fighting fish are sensitive to chlorinated water and temperature extremes. Thus, you should "age" the water in your fish bowls by filling them three-quarters full and letting them stand at room temperature for 24 hours before putting the fish in them. If you are going to purchase the bowls and fish at the same time, you can age your water ahead of time in clean milk jugs. Keep a gallon of aged water on hand at all times for cleaning the fish bowls.

   Place the bowls in a warm area of your home. (The warmer the water temperature, the more likely you are to see bubble nest building.) Set up the fish bowls as shown in Figure 4.2. Place a cardboard barrier between each bowl so that the fish will not see each other.

2. *Selecting the fish.* Purchase two male and one female Siamese fighting fish (also known as bettas) from your local pet store. Choose males with intact fins (torn fins may indicate that it has fought recently). Try to make sure that the males are aggressive by holding a small

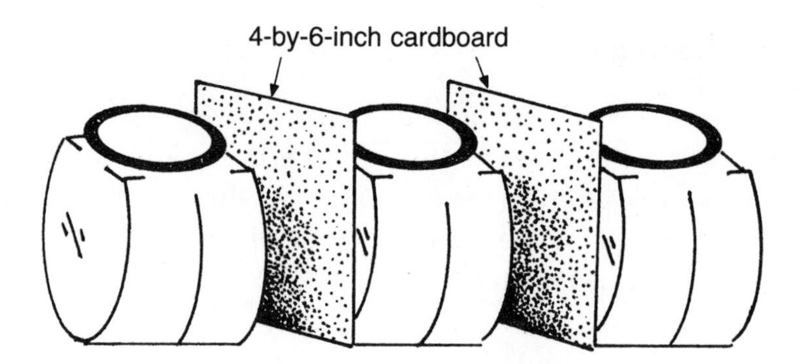

Figure 4.2. Fish bowl setup.

mirror up to their bowls. Aggressive males will display to their own mirror image (Figure 4.3). Choose a healthy looking female. One with a plump belly may be ready to lay eggs and may elicit greater display behavior. If the pet store does not carry female fighting fish, ask the store owner if he or she can order one for you.

3. *Feeding the fish.* Feed your fish a small lump of freeze-dried brine shrimp ($^1/_8$ to $^1/_4$ inch in diameter) once a day.
4. *Observing the fish.* Set aside a $2^1/_2$-hour block of time at the same time each day. You will make your observations during this time period for three consecutive days.
5. *Presenting the opponents.* Choose the most aggressive male. It will be your experimental subject. Remove the cardboard barrier and present visually three different types of opponents to your subject: another aggressive male, a female, and a mirror image. Expose the same male to each opponent for 30 minutes per day for three consecutive days, using the order shown in the table on page 37. For example purposes, it is assumed that the data are collected from 5:00

Figure 4.3. Male Siamese fighting fish displaying to mirror.

to 7:30 P.M. each day, with breaks between 5:30 and 6:00 P.M. and 6:30 and 7:00 P.M.

Time	Day 1	Day 2	Day 3
5:00 to 5:30 P.M.	Male	Mirror	Female
6:00 to 6:30 P.M.	Mirror	Female	Male
7:00 to 7:30 P.M.	Female	Male	Mirror

6. *Recording the data.* Using a data sheet like the sample shown below, record the following data:

- number of displays (extension of gill covers and fins)
- duration of each display
- number of broadside displays (bodies parallel with tail beating)
- duration of each broadside display

You may want to develop your own definition for these behaviors; for example, what constitutes the beginning of a broadside display? Just make sure that you are consistent during each of your observations. In the "Notes" section on the data sheet, record any instances you observe of bubble blowing, biting or ramming the barrier, or submissive-like behaviors.

Date: ____________
Day No.: ____________
Time: ____________

Behavior	Male	Mirror	Female
Display	6 seconds 12 seconds 8 seconds 10 seconds	5 seconds	2 seconds 10 seconds
Broadside Display	1 seconds 5 seconds 1 seconds 6 seconds	11 seconds 6 seconds	

Notes:

7. *Summarizing the data.* Count the number of displays and broadside displays for each type of opponent for each observation session. This will give you the frequency of each behavior. Then calculate the total duration of each behavior for each type of opponent for each observation session. Figure the average duration for each behavior by dividing the duration by the frequency.

Questions

Frequency, Duration, and Function of Display Behaviors

1. Which opponent received the most displays?
2. Which opponent received the most broadside displays?
3. Which opponent received the longest displays?
4. Which opponent received the longest duration of broadside displays?
5. Based on your observations, what do you think each of the display behaviors communicates?
6. Do you think the display behaviors are different for females versus male rivals?
7. Did you see any pattern in the behavior (such as alternating aggressive displays with broadside displays) that helps you understand the functions of the fish's behavior?

Display Behaviors Over Time

1. Look at your data sheets for the three days. Did the display and broadside display behaviors change over time *within* each observation session?
2. Did the display and broadside display behaviors change over time *between* the three observation sessions?
3. Were the data always consistent; for example, did one type of opponent always receive the most displays in all of the observation sessions?
4. What might you conclude about how these behaviors change over repeated encounters with a visual opponent? How did the experi-

mental design (that is, presenting opponents in different orders each day) help control for the effects of time and experience on behavior?

Other Related Behaviors

1. Did you observe any bubble blowing? If so, when did it occur? Can you relate it to the behavior of any of the opponents?
2. Did you see any biting or ramming of the barrier between opponents?
3. Did you observe any submissive behavior? If so, can you explain the circumstances under which it occurred? If you did not see any submissive behavior, why do you think that may have been the case?

How to Care for Your Siamese Fighting Fish

If you choose to keep your Siamese fighting fish, keep the males in separate bowls; otherwise, they may physically fight and tear each other's fins. The female and a male may be housed together, but they should be kept in a 5-gallon (19-l) or larger aquarium. The fish can live in bowls at room temperature and do not need a filter or air pump. Feed them daily and change their water weekly by replacing it with clean, aged tap water. You can expect your fish to live for up to a year.

CHAPTER 5

Cricket Songs

Cricket songs are one of the most well-studied forms of auditory communication. **Auditory signals** (sounds) are well suited for animal communication. They can reach great distances, can be used at night, and are easily located.

Only the adult male cricket can sing. The chirp is produced by scraping its wings. Its two front wings include a **file** (a hardened vein under one wing) and a **scraper** (a rough surface on the other wing) (Figure 5.1). When the wings are closed and rub together, a **syllable,** or sound, is produced. A typical chirp is actually a series of three to four syllables. Two to four chirps are usually repeated per second.

Studies have shown that a female listens for both the chirp rate and

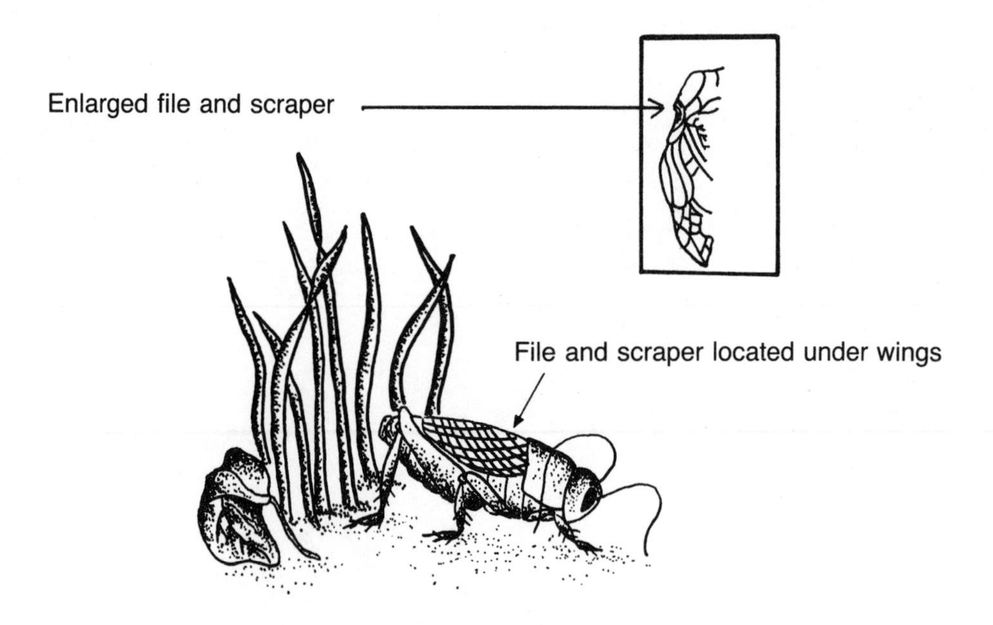

Figure 5.1. Cricket file and scraper.

syllable rate when responding to a male. Also, the louder the song, the more likely it is to attract females.

A single male may sing several different types of songs. For example, in the male domestic cricket (*Acheta domesticus*), four different songs have been identified:

Calling song. Sung to attract females ready to mate.

Courtship song. Sung during the mating process.

Aggressive song. Sung when another threatening male approaches.

Postmating song. Sung after mating has occurred.

Each species of cricket has a slightly different song. With a little practice, you can learn to identify individual cricket species through their songs.

Like fireflies, different cricket species tend to live and mate in different settings or at different times of the day. For example, one species might live in a field; yet another might inhabit the border area between the field and woods. The songs of different species may vary in their duration, rhythm, loudness, and pitch.

Purpose

To observe how crickets communicate through song.

Observational Setting

This project can be done any time of the year when you can purchase crickets at a local pet or bait store. If you are doing this project in the fall, you might want to collect your own crickets. Look under rocks and decaying logs and in the grass, or follow your ears to find them. Make sure you collect adult crickets that are all the same species.

Materials

- ☐ four 1-quart (1-l) jars
- ☐ sand
- ☐ apple
- ☐ nylon netting
- ☐ rubber bands

- ☐ live crickets (3 male; 1 female)
- ☐ three 8½-by-11-inch (22-by-28-cm) pieces of cardboard
- ☐ notebook
- ☐ pencil
- ☐ stopwatch, or watch with second hand
- ☐ 5-gallon (19-l) aquarium or equivalent container
- ☐ toilet paper tube

Procedure

1. *Making homes for the crickets.* In the bottom of each jar, place a ½ inch (2 cm) or so of moist, but not wet, sand. Keep the sand moist by adding water if necessary each day. Put a slice of apple in each jar. This will serve as the cricket's source of both food and water. Each day, remove the old apple slice and replace it with a new one. For each jar, cut a piece of nylon netting to be used for a cover for the jar, and use a rubber band to hold it in place.
2. *Purchasing or catching live crickets.* Make sure that you get three healthy adult males and one healthy female. It is relatively easy to tell the difference between females and males (Figure 5.2). The females have a long tube, called an **ovipositor,** extending from the rear of their bodies. Adult males and females have well-developed wings, whereas immature crickets simply have **wingpads.**
3. *Transferring the crickets to the jars.* Place the crickets in the refrigerator for approximately 10 minutes. The cold will slow them down and make them much easier to handle. Put one cricket in each of the four

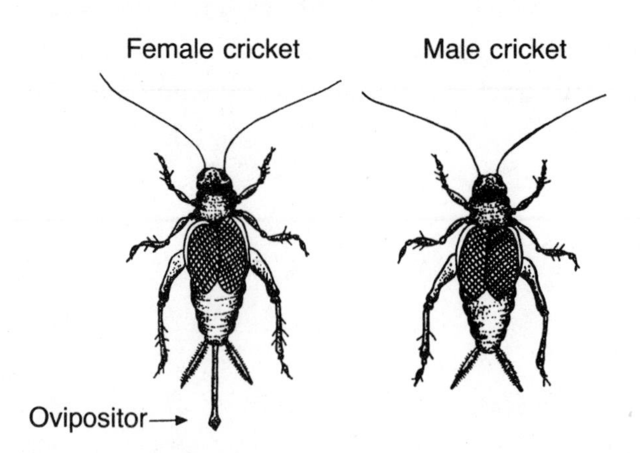

Figure 5.2. Female and male cricket.

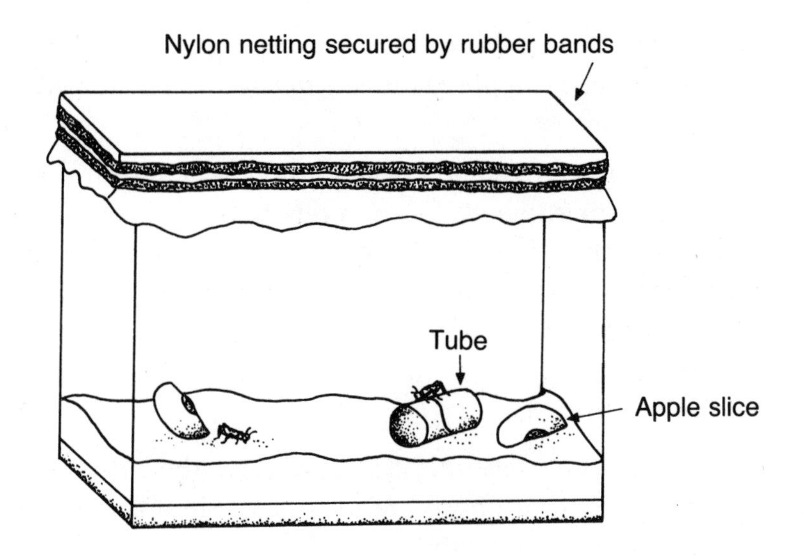

Figure 5.3. Setup for observing male–male interactions.

jars. Mark the jar containing the female. Place the nylon netting over the top of each jar, and secure it with a rubber band. Place the jars in a room that has natural light and dark. Place a cardboard divider between each of the jars so that the crickets can hear but not see each other. Check the jars every hour or so throughout the first day to determine when the crickets are most likely to sing—their active period.

NOTE: Do not let the crickets get loose in your house. They tend to eat everything, including clothes and rugs!

4. *Observing chirping.* Once you have determined the crickets' active period, make your observations for 1 hour daily during that period for five consecutive days. Record the following for each jar:

- the number of chirps per minute
- the duration of chirping
- a description of the chirping (for example, "long, short, pause, short")

NOTE: It is easiest to just make slash marks as each cricket chirps, rather than to try to count the chirps of four different crickets at once.

5. *Observing male-male interactions.* Prepare the 5-gallon (19-l) aquarium for the crickets by lining the bottom with moist sand and placing a slice of apple at each end. Place the toilet paper tube at one end of the aquarium. Cut a piece of nylon netting for the top and secure it with one or more rubber bands (Figure 5.3).

NOTE: The toilet paper tube may get slightly soggy in the damp sand. You can either use a piece of plastic tubing or place a small piece of plastic under the tube.

Introduce two male crickets into the aquarium, one at a time, waiting approximately 30 minutes before introducing the second cricket. Record the same chirping data as outlined in step 4 for 1 hour. Make detailed notes on what the two crickets do. In addition to chirping, the crickets may begin to fight. If they start fighting intensely, you might want to remove the crickets and put them back in their original jars.

6. *Observing male-female interactions.* Follow the same procedures outlined in step 5 for observing male–male interactions, but introduce the female instead of the second male.

7. *Releasing the crickets.* When you have finished your observations, let the crickets go outside if the weather is nice. Otherwise, you can keep them in their jars until the weather permits them to be released.

Questions

Chirping Activity Levels

1. After observing the crickets throughout the first day, did you identify an active period? How long was it? Was there more than one active period?

2. Can you relate anything in the environment to their periods of peak chirping? For example, is chirping related to such variables as temperature or lighting conditions?

3. Why might crickets have evolved such a sophisticated form of auditory communication rather than visual communication?

Characteristics of the Chirping

1. Did you hear each of the crickets chirp, or was it only the males that chirped?

2. How would you describe the chirp patterns you heard, in terms of rhythm, loudness, and duration?

3. Did all of the males have similar songs? If not, why might they have been different?
4. Did you hear any "dialogues" between two different crickets?
5. Did you hear any "chorusing," in which the males sang together?

Male–Male Interactions

1. Describe any vocalizations or songs that you heard. Then, compare them to the songs you heard from the individual males. Were they the same or different?
2. Did you observe any aggressive behavior? Were there any vocalizations before, during, or after aggressive encounters?
3. Did either male remain near or try to defend the tube?
4. Did you find that one male was louder than the other? Was it the more aggressive or dominant cricket that was the loudest?
5. Scientists have documented instances in which crickets use features of their environment to amplify their chirp, much like a megaphone. For example, some species use shallow burrows or leaves to amplify their sound over three times. Did you hear either male calling from within the tube? Was its chirp louder or seemingly amplified by the tube?

Male–Female Interactions

1. Describe any vocalizations or songs that you heard. Again, compare them to the songs you heard from the individual males. Were they the same or different?
2. Did you observe any sexual or aggressive behavior? Were there any vocalizations before, during, or after these encounters?
3. What was the female's response to the male's song?
4. Do you think you would get the same results with another species of cricket? If you have the time, you might expand the scope of your project by doing the same experiment with several cricket species.

PART II

Mother and Child: Projects on Mother-Infant Interactions

The bonding of mother and infant is fundamental to the survival of most species. Newly hatched or newborn animals can be classified as either **precocial** or **altricial,** depending on whether they are self-sufficient at birth. A precocial animal comes into the world with all of its sensory systems developed. An altricial animal is totally dependent on its parent for its basic needs.

Mothers of altricial young must devote a great deal of care to their offspring. Puppies, for example, are born without sight and hearing and have poor control over their body temperature. They open their eyes after approximately two weeks and begin to hear after approximately three weeks. Until this time, puppies are totally dependent on their mothers for food, warmth, and elimination activities. At five to six weeks of age, the mother begins the disengagement process, and the puppies start to focus on the outside world.

In many altricial species, caring for the young is exhausting and dangerous. For example, songbird mothers expend an incredible amount of energy maintaining a constant nest temperature. At the same time, sitting stationary on the nest increases their vulnerability to predation.

Mothers of precocial young have other issues to deal with. They must keep track of their young and protect them from predators. Most precocial species are mobile at birth or hatching and must quickly learn to follow their mother in order to survive. As a result, many species have an **inborn,** or inherited, characteristic that increases the likelihood that the young will recognize their mothers very early in life. Young ducklings, for example, are attracted to anything that moves.

Hooved mammal species have adopted several methods for ensuring that their young rest safely. Animals such as deer, gazelles, and giraffes utilize a **lying out strategy**, in which their young spend most of their first two weeks hidden under protective cover. They remain still in their hiding places for long periods of time as their mother grazes at a distance. Other hooved mammals, such as horses, sheep, rhinoceroses, and camels, utilize a **follower strategy**. The young of these species, more self-sufficient at birth, stay close to their mother when active and at rest, and separations are usually initiated by the mother.

How does the mother interact with her young at birth or hatching? What effect does this early experience have on later social development? How does the mother-infant bond change over time? Who initiates independence, the mother or the young? These are some of the questions that you can answer by observing the behavior of mothers and their offspring.

CHAPTER 6

Mother Horses and Their Young

Horses are herd animals that live together in small bands, which typically include a stallion, its mares, and their offspring. Within hours of its birth, the infant horse is standing, nursing, and beginning its socialization into the herd. By the time it is a day old, a foal can walk, trot, gallop, follow, vocalize, play with other foals, and even swim.

Foals stay very close to their mothers during the first week of their lives, usually within a distance of 15 feet (4.5 m). During this period, foals spend approximately one-half of every hour resting. Two-thirds of the time, or approximately 8 hours a day, foals rest lying down. When they sleep, they enter deep sleep and are relatively unresponsive to the events going on around them. This is a very dangerous situation for a young horse in the wild because it is subject to **predation;** that is, it is a source of food for other animals.

How does the mother horse protect her young if the foal spends a significant amount of time resting on the open ground? Recent studies have shown that the mother horse stays close to her sleeping foal by grazing in a circle around it. Usually, a grazing mare will move in a wavy-line pattern back and forth across the pasture. The mother horse also may stop grazing and rest in an upright posture with her head and neck extended over the foal.

Studies have shown that the **spatial relationship** between the mother and foal begins to change at around eight weeks of age. At this age, the foal leaves its mother more often and begins to spend more time playing. The mother's tendency to stand guard over her resting foal wanes over time, and she often grazes at distances of more than 300 feet (91 m) from where the foal is resting. The mother continues to be protective and nurses her foal until it is about a year to a year-and-a-half of age. Male foals are then driven to the periphery of the herd, and the female foals wander off when ready.

In this project you will be using a **time-sampling method** for measuring the distances maintained between mother and foal. This method assumes that the *sample* reflects a good estimate of the mother's and foal's behaviors at all other times.

Purpose

To observe how the behavioral and spatial relationships between a foal and its mother change over time.

Observational Setting

This project must be done in a farm or pasture setting in which there is a mare and her young foal. Get permission from the owner to do the observations. Also, the owner may be able to give you information on the size of the field, the horses, and so on. Other horses may or may not graze in the same pasture. It will be more interesting if you get a chance to observe the foal's interactions with other foals and adult horses. You will be observing the foal and its mother once during the first eight weeks of its life and again when it is older than eight weeks.

Materials

- ☐ measuring tape
- ☐ notebook
- ☐ pencil
- ☐ stopwatch, or watch with second hand

Procedure

1. *Mapping out your observational setting.* To estimate the distances between the mother and her foal accurately, you need to determine the size of the pasture and to designate some landmarks as guides. Use the measuring tape to estimate the length of your gait (the distance from the toe on one foot to the heel on the other foot as you walk). Then, estimate the dimensions of the pasture by pacing off the length and width of the field.

   Locate easily identified landmarks, such as fence posts, trees, and large rocks. Measure the distances between all the various landmarks. Also, ask the owner if you can measure the length of both the foal and its mother to use as guides. Once you have made all of these measurements, draw a to-scale map (Figure 6.1). Refer to this map to estimate distances as you observe mother–young interactions.

2. *Observing the spatial relationship of the foal and its mother.* Before the foal is eight weeks old, observe the mother-foal pair for a total of 10 hours at various times of the day. This can be done over several days, but each observation period should be at least 2 hours long. Every 5 minutes, estimate and record the distance between the mother and her foal. Use a data sheet like the sample shown on the following page to record your information. After 10 hours of observation, you will have 120 distance estimates.

   Repeat your observations again for 10 hours when the foal is eight weeks or older. You will have an additional 120 distance estimates for comparison.

3. *Observing the behavioral state of the mother and foal.* As you record the

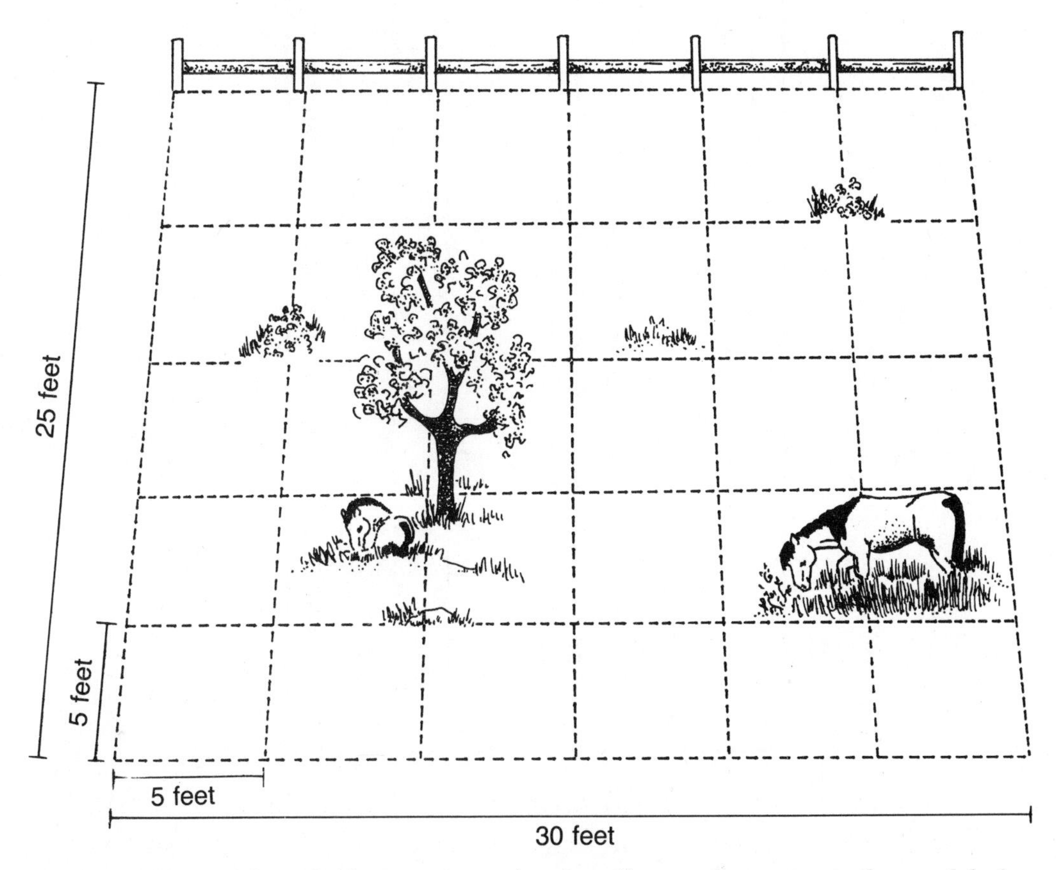

Figure 6.1. Grid overlaid on pasture showing distance between mother and foal.

Mother–Foal Interactions

Date: ______________
Begin Time: ______________

Scan	Distance	Rest Down	Rest Up	Feeding	Nursing	Playing	Active	Approach	Retreat
11:05 A.M.	10 feet	M	I						
11:10 A.M.	<1 foot				I			I	
11:15 A.M.									
11:20 A.M.									

Notes: Use **M** to indicate *mother* and **I** to indicate *infant*.

distance estimate between the mother and her foal, also record the behavioral state of each, as follows:

- resting lying down
- resting standing up
- grazing/feeding/drinking
- nursing
- playing (alone or with other horses)
- active (alert, but not doing preceding behaviors)

4. *Observing mother-foal interactions.* After scanning the pasture to estimate distances and record behavioral states, observe the interactions of the mother and her foal. Record instances of the following behaviors, indicating whether the mother or the foal was the initiator:

Approach. Mother moves toward her foal or the foal moves toward its mother.

Retreat. Mother moves away from her foal or the foal moves away from its mother.

Questions

Spatial Relationship Between Mother and Foal

1. Calculate the average distance maintained between the foal and mother for the pre-eight-week observation session and compare the averages for the pre- and post-eight-week observations. Does the foal stay closer to its mother when it is younger?
2. Look at the range (closest distance recorded and the farthest distance recorded) for each of the observation sessions. Was the range higher as the foal got older?

Behavioral State of the Mother and Foal

1. Calculate the percentage of time that the foal and its mother spend at each activity for the two observation sessions. Sum the number of times you recorded a behavioral state and divide it by the total number of observations (120). For example, if the foal was resting lying down during 28 of the scans, you can estimate that it spends 23% of its time lying down.

2. Compare the activities of the foal and its mother for the two observation sessions. Do you see significant changes in the activities of the foal over time? Why might you see these changes?
3. Do you see significant changes in the activities of the mother over time? Are they consistent with changes in the foal's behavior?
4. Do you think you would get different results if you observed the foal again at about six months of age? Why?

Mother–Foal Interactions

1. Who is responsible for the foal's increasing independence, the mother or the foal? Does the mother engage in behaviors that train the foal to be more independent? Or does the foal's following behavior wane over time?
2. Do you think that the foal is moving away from its mother or toward other experiences and members of the herd?
3. Why is it to the foal's advantage to become more independent as quickly as possible?
4. Do you think that the behavior of domestic horses would be similar to the behavior of wild horses? Why?

CHAPTER 7

The Following Behavior of Mallard Ducklings

Young precocial animals (animals that are born with all of their sensory systems developed) learn to recognize their mothers very early in life. Many of these animals demonstrate a tendency to follow or approach a moving object in the first few hours after hatching or birth. This behavior is known as **imprinting.** It occurs in species such as ducks, chickens, shrews, and sheep. As mobile newborns, these animals must quickly learn to follow their mother and go to her in order to survive.

The approach and following response of young, precocial birds has been thoroughly studied in the laboratory. Within a few hours after hatching, mallard ducklings have followed duck decoys, people, a football, a matchbox, and other objects resembling their parents. Movement of the object is important to the development of the duckling's following response. Jerky, uneven movements, resembling more the natural behavior of a mother duck, have been shown to be more effective in eliciting imprinting than constant, continuous movements.

Mallard ducks, which nest near the water in open fields or tall grasses, see their mother move away from the nest right after they hatch and must follow her. Thus, mallard ducks must be highly responsive to moving objects.

Wild mallard ducks are found throughout the Northern Hemisphere. They live and nest near the water and are skillful and energetic swimmers; the ducklings start swimming the day they hatch. Typically, a mallard duck's territory includes a body of water for feeding, tall grasses or reeds for protection, and a **loafing area** for resting and preening (Figure 7.1).

Nest building typically occurs during a two-week period in March and April. Because mallards are quite secretive during the breeding

Figure 7.1. Mallard duck territory.

season and hide their nests well, it is often impossible to observe their nests. However, during nest building the female mallard quacks constantly (male mallards do not quack but, rather, make a nasal whistling sound). The average incubation period of the eggs is 23 days, and the ducklings often leave the nest the same day they hatch. In mallard ducks, the optimal period for forming attachments is 25 to 50 hours after hatching.

At first the female **broods** the ducklings in the loafing area by covering them with her wings, but as they get larger they sleep next to her. From the day they hatch, the ducklings follow their mother, and she keeps a constant watch over them. Because of the position of her eyes on her head, the mother duck can see nearly 360 degrees without even moving her head. By 50 to 60 days after hatching, the ducklings **fledge** (start to fly) and are left on their own by their mother, who then goes off to **molt** (shed her feathers).

Purpose

To observe the following response of mallard ducklings.

Observational Setting

Mallard ducks can be best observed on ponds and lakes in city parks, suburban parks, and wildlife preserves and on natural ponds and lakes in rural areas. Some people also breed and raise domestic mallard ducks. In most areas, to observe mallard ducklings, you must look for them in April and May.

Materials

- ☐ notebook
- ☐ pencil
- ☐ stopwatch, or watch with second hand
- ☐ stale bread/bread crusts

Procedure

1. *Identifying the ducks.* Male mallard ducks may be identified by their distinctive coloring: an iridescent green head with a white neck band, reddish brown breast, mostly white tail feathers, and black and dark gray wings. The females lack the green head and are streaked brown all over. In early to midsummer, the male undergoes a partial molt; for one month it looks exactly like the female.
2. *Locating a mother duck and her ducklings.* After you have located a population of mallard ducks to observe, make daily trips to the area to see if you can identify a nesting female. During the nesting phase, the female will be identifiable by her persistent quacking. Approximately one month after the nesting phase begins, you can expect the eggs to hatch. It is not essential that you observe the ducklings from their very first days, but you will want to make your first observation during their first two to three weeks of age.
3. *Observing the mother and her ducklings.* Observe how the ducklings follow their mother in their first few weeks of life (Figure 7.2). Using natural landmarks (for example, trees, reeds, shore points), estimate the average distance maintained between the mother and individual ducklings. Watch them swim around and feed for 1 hour. Every 10 minutes, make another estimation of the distances between the mother and the individual ducklings.

Figure 7.2. Following behavior of mallard ducklings.

4. *Observing the mother and her ducklings during feeding*. Choose an area of water close to shore that you can estimate the size of. Throw some bread crumbs into that area, trying to spread them as uniformly as possible. Watch the behavior of the mother and her ducklings. Estimate again the average distances maintained between individuals. Repeat the estimation every 10 minutes until all of the food is gone. Note the presence of any other ducks as well as the mother and ducklings' response to them. Watch and note how long it takes the mother and her ducklings to regroup.
5. Repeat steps 3 and 4 at least one time, two to four weeks later. (Ideally, a third observation, near the time the ducklings first fledge, at approximately 50 to 60 days of age, would produce even more interesting results.)

Questions

Early Following Behavior

1. Do all of the ducklings follow the mother and swim in the same direction? Calculate the average distance maintained between the mother and her ducklings for the first hour of observation. Is it relatively constant over the hour or does it vary?
2. Does the mother ever have to retrieve any of her ducklings? Does

she vocalize? Do her ducklings vocalize? How does the mother duck bring a stray duckling back into the group?

3. How does the presence of food affect the distances maintained between the mother and her ducklings? How long does it take the ducks to regroup after the food is gone? Does the mother indicate in some way that the ducklings should regroup?
4. How do the mother and her ducklings react to the presence of other ducks?

Later Following Behavior

1. How does the ducklings' following behavior change over time? Calculate the average distance maintained during the second observation session. How does it compare to your earlier observations?
2. Do the ducklings still swim in the same direction as their mother? Does the mother ever swim away from the ducklings?
3. Does the average distance maintained between the mother and her ducklings change when food is present? Are the distances now greater than they were when the ducklings were younger? Do the ducks still regroup after feeding? How has the behavior changed from earlier observations?
4. Do the mother and her ducklings react to the presence of other ducks the same way?

Behavior at Separation

1. Young mallard ducklings fledge at approximately 50 to 60 days of age. At this time, the mother goes off to molt. Do you think that the separation is abrupt or that the ducklings get increasingly independent?
2. Based on your observations, who do you think initiates the separation or independence? The mother? The ducklings?
3. Why is the following response so important at the time of hatching? Why might it be dangerous for a duckling to continue following behavior as it grows older and more independent?

CHAPTER 8

Smiling in Human Infants

Smiling is the most important facial expression when it comes to human bonding. In fact, scientists have identified 180 forms of human smiling that are physically and visually distinct. We smile as an expression of many different emotions: for joy, in amusement, in appreciation, and in fear. We smile when we greet one another, we smile in sympathy, and we smile in apology.

No one is quite sure how smiling develops, but it appears to involve both maturation and learning. By around three weeks of age, an infant will smile at the sound of a human voice. By four to six weeks, an infant will smile in response to a number of things, including faces, interesting visual patterns, and pictures of one, two, or many eyes. By the time an infant is two to three months old, it can recognize specific faces. From this point on, it smiles most in response to familiar individuals.

Why is smiling so significant and why might an infant start smiling at birth? The infant's smile is thought to be a very important social signal that bonds the mother and child. The infant uses crying to attract the mother's attention. It uses smiling to keep her close. As the infant begins social smiling, the amount of time a parent spends with it increases significantly. Infants reared in family situations tend to engage in social smiling earlier than those reared in institutions. In addition, social smiling has been shown to be delayed by mother-infant interactions described as hostile or cold.

Evidence that smiling is critical to mother-infant bonding is supported by findings that women tend to be more attuned to infant smiling. For example, women are better than men at remembering smiling faces of infants. However, there is no difference in the memory of men and women of infant faces showing other facial expressions.

Infants tend to use different smiles for different situations (Figure 8.1). They smile differently at their mother versus at a stranger, and they smile differently at patterns of dots versus at people. Researchers be-

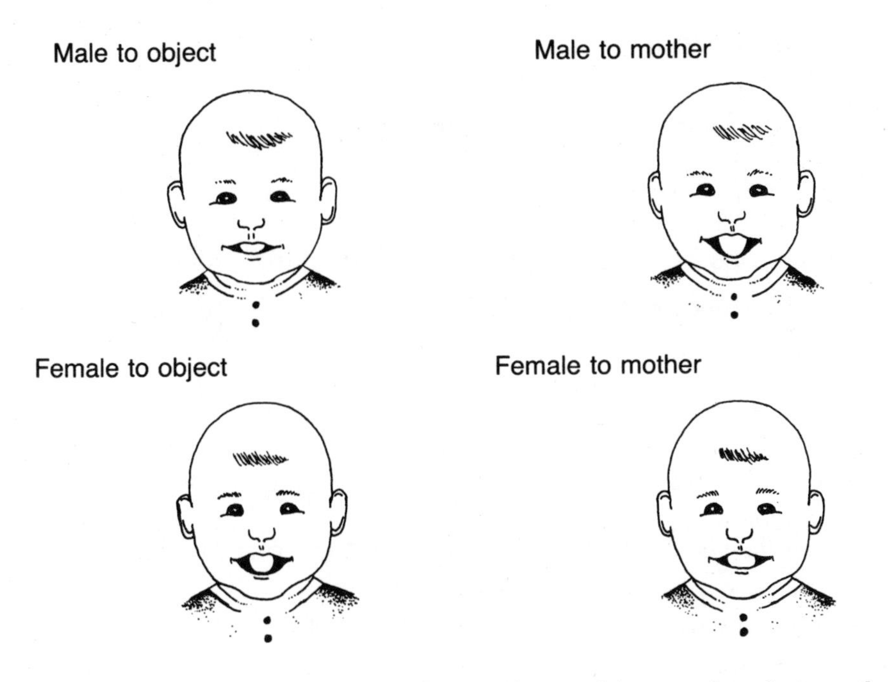

Figure 8.1. Male and female infant smiles to objects and to their mothers.

lieve that infants develop at least four different smiles by the time they are eight weeks to twelve months of age.

Perhaps the most interesting finding is that infant boys tend to smile differently than girls by eight weeks of age. Male and female infants smile differently when viewing a picture, an object, or another person. For example, when a male infant smiles at his mother, his eyebrows are raised and more vertical; his mouth is wide open and rounded, with the corners turned up. A female infant smiles with rounder eyes; her mouth is also wide, but her lips are only partially separated. Studies also have shown that nurses who work with infants can detect sex differences in newborns on the basis of facial cues alone.

Are women better than men at telling infant boys from girls on the basis of their smile? Does experience affect one's ability to tell the difference between the two? Are pregnant women and new mothers better able to tell infants apart on the basis of their smiles? These are some of the questions you will be asking yourself as you perform the following experiment.

Purpose

To determine whether individuals can identify the sex of an infant based on its smile.

Observational Setting

This project can be done in any setting.

Materials

- ☐ 5 photographs of 8-week to 12-month-old smiling male infants
- ☐ 5 photographs of 8-week to 12-month-old smiling female infants
- ☐ notebook
- ☐ pencil

Procedure

1. *Collecting the photographs.* Collect five photographs of smiling infant boys and five photographs of smiling infant girls. The infants all should be approximately 8 weeks to 12 months of age. These photographs may be your baby pictures, pictures of siblings or relatives, pictures that you or someone else took, or pictures obtained from friends. It is important that the photographs contain no hints as to the sex of the infant. All photographs should follow these basic rules:
   - The infant should be clearly smiling.
   - The infant should be wearing gender-neutral clothes.
   - No gender-specific toys should be visible in the picture.
   - No gender-specific bedding should be visible in the picture.
   - Clothing or background colors should be gender-neutral (for example, stay away from color stereotypes such as blue and pink).
2. *Preparing the photographs.* Shuffle the stack of photographs so they are in no specific order. Then, number each of the photographs on the back using a number from one to ten. Test them out on several friends. First, ask them to identify the sex of each infant. Then, go through the pictures, one by one, with them and ask them the reasons behind their choices. This will help you identify whether any of the photographs still contain gender-specific cues. Replace any photographs that contain gender-specific cues.
3. *Finding a group of subjects.* Select five males and five females to be your subjects. You may choose any age group, from adolescent to adult, but make sure that **all** of your subjects are similar ages. Also, find out whether or not they have any experience with infants (for example, a mother, a babysitter).

Figure 8.2. Sample photographs of smiling infants.

4. *Testing your subjects.* Seat each subject at a table with no distractions. Before presenting them with the photographs, give them a blank data sheet like the following one.

Subject Name: ____________
Subject Sex: ____________
Date: ____________

Photograph No.	Infant's Sex
1	______
2	______
3	______
4	______
5	______
6	______
7	______
8	______
9	______
10	______

Then, tell them the following:

> *This is an experiment about how individuals recognize the difference between male and female infants. Please indicate whether you think the infant in each photograph is a male or a female by writing an M or F, respectively, next to the photograph number.*

Do not tell them that this is an experiment about smiling or facial expression because it may bias your results. Shuffle the stack of photographs before presenting them to each subject. Do not provide any feedback on the correct answers, even if asked. Ask your subjects not to discuss the experiment with their friends. When you are done collecting data from all ten subjects, you then may discuss the purpose of the experiment and the results.

Questions

Overall Analysis of the Data

1. From your ten subjects, who each viewed ten photographs, you have 100 responses. Calculate the overall percentage of correct responses by adding up the number of correct responses and dividing by 100 (the total number of responses). This will tell you whether individuals can identify the sex of the infants at better-than-chance levels. If your overall percentage is 50% or less, you may conclude that the overall recognition ability is less than or equal to chance. If the overall percentage is greater than 50%, your subjects are showing some ability to recognize the sex of infants based on their smiles.
2. What do your overall results indicate? Did the individuals correctly identify the sex of infants at better-than-chance levels? How might you explain your results?

Analysis by Sex of the Subject

1. Does the sex of the subject affect one's ability to correctly identify the photographs? Separate your data sheets according to the sex of the subject. Calculate the percentage of correct responses for female subjects only. Divide the number of correct responses by 50 (the total number of responses by female subjects). Then, do the same thing for the male subjects. Do your results indicate that either males or females are correctly identifying the photographs at better-than-chance levels?
2. How would you explain your results?

Analysis by Sex of the Infant

1. Do you think that male infants may be easier to identify than female infants or vice versa? Combine all of your data sheets and calculate the percentage of correct responses given for the male infant photographs and the percentage of correct responses for the female infant photographs. Divide the number of correct responses for female infants by 50. Then, do the same thing for the male infants. What do your results show?
2. How would you explain your results?

Photograph-by-Photograph Analysis

1. Are some photographs, regardless of the sex of the infant, more often identified correctly? With your combined data sheets, tally the number of correct responses for each photograph (use the photograph number to aid in your analysis). Calculate the percentage of correct responses for each photograph by dividing the number of correct responses by 10 (the total number of responses for each photograph). Are certain faces easily identified correctly? Are other faces difficult to identify?
2. Examine each of the photographs and try to summarize why you think some faces are easier to identify than others. Are there common facial features, similar facial expressions, or other matching qualities? Review the results of your other analyses. Does there appear to be a pattern in how these faces are perceived?

CHAPTER 9

Lambs and Mothers

All of a lamb's sensory systems are functional at birth, and in some breeds, lambs stand within minutes of birth. If the mother is not responsive, competing objects can lead the lamb away from her. For example, it is not uncommon for a firstborn lamb to stray while the mother is licking and tending to its twin.

Typically, a ewe's first response after birth is to rise and begin licking her lamb, whereby she becomes familiar with its scent. She continues to make soft, low-pitched bleats or rumbles, which began during the birthing process. The lamb raises and shakes its head. It then rights itself, raises up on its knees, and stands. Nudging and nuzzling its mother, it moves toward her udder.

The mother responds in one of two ways: She either prevents her lamb from nursing or assumes a **receptive posture.** She prevents the lamb from nursing by moving her hindquarters away with a circling movement. If she is done licking the lamb and is ready to nurse, she assumes a receptive posture, an inverse-parallel position (lamb's head near its mother's tail), arching her back and extending her hind leg to raise her teat (Figure 9.1).

The combined effects of the mother's presence, her licking, and her low-frequency vocalizations have a calming effect on the lamb. In the absence of these behaviors, the lamb is quick to stand. It will head for the nearest large object and attempt to place its head under any protruding thing. Thus, the interaction of the ewe and the lamb in the first minutes after birth is critical to the development of the mother-young bond.

Maternal behavior has been shown to develop more slowly in inexperienced ewes. Sometimes an inexperienced ewe will run away from or kick her newborn lamb. Or she may spend a great deal of time circling the lamb, preventing it from nursing. Usually within 3 hours, the ewe begins to care for the lamb.

Figure 9.1. Lamb nursing in an inverse-parallel position.

A number of researchers have tried to isolate the factors responsible for the development of **maternal behavior** in ewes. Studies have shown that stimulation of the birth canal (through the birthing process), the smell of the **amniotic fluid,** and the physical presence of the lamb are all essential to the development of normal maternal behavior in inexperienced ewes. Sometimes even with this stimulation, it takes a while for the new mother to learn how to handle her newborn lamb.

Purpose

To observe the differences in maternal behavior exhibited by inexperienced and experienced sheep mothers.

Observational Setting

This project must be done in a setting in which you can observe lambs just after birth. The ideal setting is a sheep farm in spring. Many lambs may be born over several weeks. It also may be possible to make arrangements to observe lambs with someone that has a small herd. This might be a local farmer or a petting zoo.

Sheep breeders watch the development of the pregnant ewe's udder for clues to the stage of her pregnancy. When the udder becomes pink and congested, lambing is imminent. Make an arrangement with the sheep owner to call you when the ewe is about to give birth.

You will be observing the behavior of inexperienced and experienced ewes and their lambs in the first few hours after birth. Make sure that you ask the sheep owner to tell you whether or not this is the ewe's first birth.

NOTE: If you are not able to observe the sheep just after birth, you may do this project with several-day-old lambs. Record all of the appropriate behaviors listed in the Procedure.

Materials

- ☐ notebook
- ☐ pencil
- ☐ stopwatch, or watch with second hand

Procedure

1. *Getting into position.* If possible, arrive at the birthing area just as the ewe is about to give birth. If it is a first-time mother, the owner may ask you to stand away from the lambing pen or birth area. The behaviors you need to observe, standing, licking, nursing, and so on, can be easily viewed from a distance. Position yourself so that you can observe behaviors without frightening the ewe or getting in the way of people assisting with the birth.
2. *Recording the behaviors of the ewe.* Note the time at which the birth occurs. Observe the ewe and lamb for a 2-hour period and record the following behaviors of the ewe:

   Time to rise. The number of minutes and seconds it takes for the mother to rise after the lamb is born.

   Frequency of vocalizations. Record the number of vocalizations made by the ewe.

   Time to lick. The number of minutes and seconds it takes for the mother to begin licking her lamb after it is born.

   Duration of licking. The amount of time in minutes and seconds the mother spends licking her lamb. Record the duration of each licking bout.

   Time to assume nursing posture. The number of minutes it takes for the mother to maneuver herself into a receptive posture with the lamb. The nursing posture, arching her back and lifting her hind leg, should follow shortly thereafter.

   Instances of circling. Number of times the mother moves her hind-quarters away from the lamb and nudges the lamb back toward her head.

Figure 9.2. Newborn lamb trying to stand.

Instances of retreat. Number of times and duration of time that the mother spends walking or turning away from the lamb.

3. *Recording the behaviors of the lamb.* While recording the ewe's behavior, record the following behaviors of the lamb:

Time to stand. The number of minutes and seconds it takes for the lamb to rise after it is born (Figure 9.2).

Frequency of vocalizations. The number of vocalizations made by the lamb and the behavior of the mother at the time the lamb vocalizes.

Attempts to nurse. The number of times the lamb nudges or nuzzles the mother in the udder region.

Duration of nursing. The number of minutes and seconds the lamb spends nursing.

NOTE: The ewe may have more than one lamb. Although it may be hectic, try to record the behavior of each lamb. Record what the other lamb is doing while one is nursing or being licked by the mother.

4. *Making additional observations.* Repeat your observations for as many experienced and inexperienced mothers as you can. You also may want to return to observe the sheep and their lambs several weeks later. It would be interesting to observe and record the same behaviors after the mothers are more experienced and the lambs are more robust.

5. *Organizing your data.* Organize your data into a table like the one on page 70 to compare the behaviors of experienced and inexperienced mothers.

	Experienced Mothers	Inexperienced Mothers
Ewe Behaviors		
Time to rise Frequency of vocalizations Time to lick Duration of licking Time to assume nursing posture Instances of circling Instances of retreat		
Lamb Behaviors		
Time to stand Frequency of vocalizations Attempts to nurse Duration of nursing		

Questions

Behavior of Experienced versus Inexperienced Mothers

1. Calculate the average time to respond, duration, and number of instances of behaviors observed (for example, the average number of times ewe mothers circled) for the experienced and inexperienced mothers. For example, if you observed four experienced mothers, calculate their average time to rise by adding the four time to rise numbers you recorded and dividing by 4 (the number of experienced mothers). This will enable you to make some quantitative comparisons between the behavior of experienced and inexperienced mothers.
2. Did you find behavioral differences between the experienced and inexperienced mothers? What behaviors would you have expected to be different?
3. Did you observe instances of circling behavior or actual retreats? Did these occur with the experienced mothers as well as the inexperienced mothers?
4. Why might an inexperienced mother lick her lamb longer than an experienced one?

Behavior of the Lambs

1. Did the behavior of the lambs of experienced and inexperienced mothers differ?
2. Why might it be to a lamb's advantage to stand sooner if its mother is less responsive?
3. Did either group of lambs seem more independent (for example, moving away from the mother)?
4. Do you think you would see the same differences in the lambs if you observed them a day later? Two weeks later?

CHAPTER 10

Songbird Parents: A Case Study with Bluejays

Songbirds must invest a lot of time and energy in the **incubation** and care of their offspring. To incubate its eggs, a songbird mother must maintain a high, even temperature. This means sitting on the eggs almost constantly. How does the female eat and still have enough energy left to warm its eggs? In the case of the bluejay, the male at first feeds her in the vicinity of the nest. In fact, this is one of the only ways to tell the male from the female. During the latter stages of incubation, the male brings food to her nest. The periods of incubation become longer and longer until the female rarely leaves her nest. During this time, the male usually stays near the nest except to feed.

Bluejays typically have a **clutch** of four to five eggs and an incubation period of 17 days. After hatching, the female bluejay sits on, or broods, the **nestlings** for the first few days. Both the male and the female feed the young, and the male continues to bring food for the female. The sight of the babies' gaping mouths and the begging calls signal to the parents that the young need to feed.

During incubation and the **nestling phase,** the songbird family is most exposed to predation. This is because of the increased activity around the nest as well as potential signs that the nest is inhabited. The songbirds must be careful to hide the fact that there are eggs or young in the nest. For example, after hatching parents must remove empty eggshells without calling attention to the nest.

How do songbird babies stimulate their parents to feed them without drawing the attention of potential predators? Studies have shown that songbirds can visually discriminate between their parents and other objects in the environment. Typically, they gape at the sight of their parents and crouch at the sight of strange objects. The larger the songbird, the later the crouching response develops. Crouching usually begins at days 5 to 7 in smaller songbirds, at day 14 for bluejays, and at day 18 for crows.

It is also interesting that bluejay parents care for their young longer than most songbirds. Many small songbird babies are independent at 25 to 30 days, or two to three weeks after leaving the nest. Bluejay parents, in contrast, tend to their young for up to four months. The family group also stays together for some time after the young leave the nest. It is not unusual to see the family together well into fall.

Purpose

To observe the roles that the mother and father play in the incubation and care of bluejay young.

Observational Setting

Bluejays are typically found in the eastern and central United States, south to Florida and Texas. If you live in another area of the country, you may choose to look at another member of the jay family. Breeding usually occurs from May through July, so plan on making your observations during the summer months.

Bluejays are fairly common birds, but you might have to do a little sleuthing to find their nest. This is because mated pairs tend to be very quiet and remain close to their future nest site. Otherwise, bluejays can be very raucous and easy to spot.

Look for their nests in areas of mixed scrub and mature growth. They typically nest in evergreen trees, close to the trunk, and approximately 10 to 20 feet (3 to 7 m) from the ground. You may have to climb a nearby tree to get a good view of the nest. Bluejay nests have a crude appearance, with the outside made of sticks and twigs. They are usually lined with bark strips, grasses, moss, and fine roots. Sometimes you might find artificial materials such as plastic, rope, twine, and paper woven into the nest.

Be careful not to get too close to the nest, for your scent alone can attract predators. If possible, try to find a nearby tree that you can climb or in which you can establish an observation post.

Materials

- ☐ notebook
- ☐ pencil

- ☐ stopwatch, or watch with second hand
- ☐ binoculars (7.5 × 35 power), if available
- ☐ bird identification guide (if watching birds other than bluejays)

Procedure

1. *Identifying a bluejay.* Locate a pair of bluejays (Figure 10.1). The male and female look identical. A bluejay is a large songbird with blue wings and tail. It has a prominent gray-blue crest. A bluejay's wings have white and black areas in them, and its blue tail has white tips. Its underparts are white-gray. Its vocalizations include whistles, warbles, and screams.
2. *Observing the parents at the nest.* Observe the nest for several hours at around the same time each day. Note the following:
   - the duration of time the female sits on the nest.
   - the duration of time the female is off the nest.

Figure 10.1. Female bluejay on nest with male nearby.

Figure 10.2. Nest of baby bluejays.

- the presence of the male near the nest.
- approximately how many feet (meters) the male is from the nest.
- any instances of the male feeding the female, including when and where it occurred.

3. *Counting the baby birds.* Watch the nest, or check it daily until the eggs hatch, and count the number of nestlings.

4. *Observing the parents and the young.* Make the same observations shown in step 2 until the babies fledge, with the addition of the following:
   - when the young are fed and by whom (Figure 10.2).
   - what, if anything, is removed from nest.

Questions

Observing the Mother

1. How long did you observe the female before the eggs hatched?

2. Just before the eggs hatched, you probably noticed that the female sat on the eggs for longer periods. Why do you think the female spends more time at the nest just before the eggs hatch?

Observing the Father

1. Did the male always stay near the nest?
2. Were you able to observe the male feeding the female? Did it occur at the nest site?
3. Why do you think the male feeds the female? Why do you think the male does not take a turn sitting on the nest instead?

Observing the Eggs

1. Did all of the eggs hatch? If not, what happened to the egg(s) that did not hatch?
2. Why do you think that some birds have more eggs than others?
3. In some songbird species, the male does not assist the female in caring for the young. Would you expect these birds to have more or fewer eggs than bluejays?
4. Did you observe the male or the female carrying any material (for example, eggshells, droppings) away from the nest?

Observing the Baby Birds

1. Did you ever hear the baby birds begging for food? If so, how long could you hear them?
2. Did either the male or female always stay near the nest with the young?
3. How did the parents care for the young? After the eggs hatched, were you still able to tell the male from the female parent? Did they perform some of the same roles?
4. What happened just before the baby birds were fed? Did they make sounds? Could you see them gaping their bills?
5. Did you ever see the baby birds crouch? What may have happened that made them crouch?
6. Did you ever hear the parents call out? What was the response of the baby birds?
7. If you observed the baby birds leaving the nest, did you see both parents continue to feed them?
8. Why do you suppose that bluejay parents stay with their young longer than most smaller songbirds?

PART III

Play and Social Behavior: Projects on Animal Social Interactions

Animals have many forms of social groups. There are flocks of birds, herds of elephants, troops of baboons, prides of lions, schools of fish, clusters of butterflies, colonies of termites, and human families, to name a few. Some animals remain in social groups throughout their lives. Others are solitary except when mating.

Within social groups, animals may have very different roles. Most social groups have a leader. This leader may be either a male or female. For example, an elephant herd is typically a group of up to 20 females. The herd is led by the oldest and strongest female. The other females in the group are usually her daughters and granddaughters. In packs of wolves, there are two leaders. One male is dominant, and one female is dominant. The dominant male hunts for food and defends the feeding areas. The dominant female is the only one that has offspring, even though there are other mature females in the pack.

Some groups of animals have a definite social order. Each member of the group has its own place or rank. In a flock of hens, for example, a pecking order is established. The top hen can peck all other hens without being pecked back. The second-ranking hen can peck all hens except the top-ranking hen. At the bottom is a hen that is pecked by every hen but does not peck back. The higher a hen is in the pecking order, the better access it has to food and choice nesting places.

Social behavior in many animals begins as play behavior. For example, the pack behavior of dogs is developed during a socialization period, which lasts from approximately 4 to 12 weeks of age. During this time, puppies use play behavior to learn about their litter mates and the outside world. By 14 weeks, puppies have a clear sense of their social status.

As you observe groups of animals, try to understand the relationships between animals. Are you observing the entire group? Are males, females, and young present? Is there a dominant animal? Is the dominant animal male or female? Is there some form of social order? Is the male, female, or both caring for the young? Is it mating season? Each of the projects that follow focuses on one form of social interaction. There are many more for you to observe and discover on your own.

CHAPTER 11

Neon Tetra Schooling Behavior

Over 4,000 kinds of fish swim together with their own kind in groups known as **schools.** Schools vary in size from several fish to billions. Schools may be miles long and wide. The fish that school may be small, like herring, or large, like the 1,800-pound (817-kg) bluefin tuna. The fish in an individual school are close to the same size. They swim in the same direction and at a constant pace.

Schools of fish appear to form spontaneously, yet they do not follow any leader or set of leaders. When the group changes direction, those fish on the flank, or the outside of the school, simply become the leading edge of the school. Typically, the group behaves as though it were a single fish.

Why do fish school? One theory is that fish in groups are better protected from threats such as larger fish that eat them. By congregating in schools, they limit the areas of the ocean where they can be found and attacked. Others believe that schools exist because there is safety in numbers. A group of many fish can confuse a larger fish and prevent it from focusing on a single fish. Another theory suggests that a number of little fish together may look like a larger fish, so they are less likely to be eaten.

How do the fish stay together and all swim in the same direction? How do they come back together when they split apart? Experiments have shown that fish use a number of their senses to maintain schools. These include vision, touch (sensitivity to changes in water pressure), and chemical signals through the water. In the following experiment, you will be testing both the visual and chemical communication of neon tetra fish.

Purpose

To observe how and why fish school.

Observational Setting

This project can be conducted in your home if you have an aquarium that is 10 gallons (38 l) or larger. It also can be done in a pet store with the cooperation of the owner.

Materials

- ☐ 10-gallon (38-l) aquarium
- ☐ 3 or more neon tetra fish
- ☐ fish food
- ☐ small fish net
- ☐ plastic opaque container, such as a margarine tub, with 50 to 100 small holes in its sides and bottom to permit water transfer
- ☐ optional: 1 larger, more aggressive fish such as a tiger barb

Procedure

1. *Observing the school.* Observe the school of fish by themselves in the tank for 10 minutes. Note the distances between fish and the shape of the school. Do they always stay in a schooling pattern? What makes them change direction? Drop a small amount of fish food in the aquarium (Figure 11.1). Observe what happens to the size and shape of the school while they are eating. When the food is gone, gently lower the net into the aquarium and move it toward the school. What happens when the net gets closer to the fish? (Option: Introduce a larger, more aggressive fish. Observe what happens to the school when the larger fish approaches.)

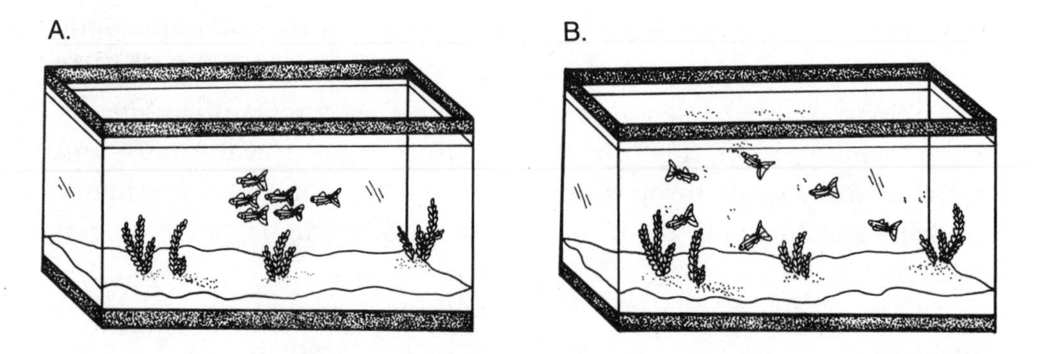

Figure 11.1. Characteristics of school of fish before (A) and after (B) feeding.

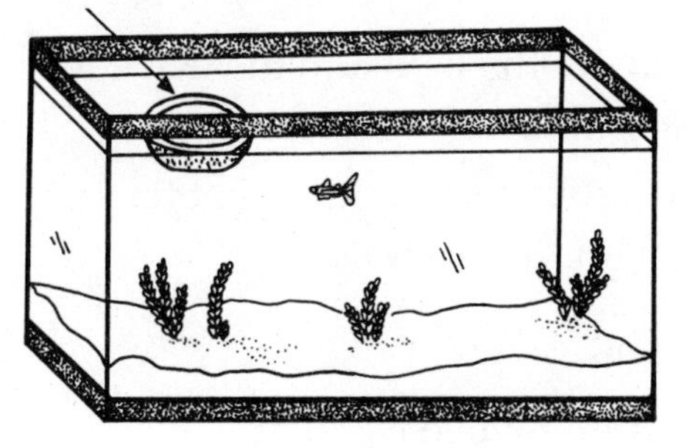

Figure 11.2. Aquarium setup for testing chemical communication.

2. *Testing for visual communication.* Turn off all the lights so that you can barely see the fish. Observe the school of fish by themselves in the tank for 10 minutes. Note what happens to the school as the fish get used to the dark.
3. *Testing for chemical communication.* Remove all but one neon tetra from the tank. Observe where it swims for the next 5 minutes. Lower the container into the aquarium, and put all of the remaining neon tetras in the container. Make sure that the top of the container is slightly above the surface of the water so that the fish cannot swim out (Figure 11.2). Observe where the single neon tetra swims for the next 10 minutes. Does it swim randomly in the water? Does it swim around the container?

Questions

Observing Schooling Behavior

1. Compare the schooling behavior of the fish before you fed them to their behavior while they were eating. Did you see the shape of the school change? Did the distances between fish increase as they searched for pieces of food? Do you think that schooling behavior increases or decreases the efficiency of their feeding?
2. What happened when you introduced a threat (either the net or the larger, aggressive fish)? Did the fish group tightly together, or did they swim in all directions? As the net approached the school, did the school split in half? Where did they swim? Where and how did they rejoin each other? Did you observe what might be described as a defensive strategy? Did it work well as a strategy?

Visual Communication

1. When you turned out the light and the fish got used to the dark, did the shape of the school change?
2. Based on your results, do you think that the fish use visual communication to maintain their school?

Chemical Communication

1. Compare the swimming behavior of the individual neon tetra before and after the container of fish was placed in the aquarium. What was the initial response of the individual fish? What did it do after a while? Did it approach the container?
2. Based on your results, do you think that the fish use chemical signals to maintain their school?

CHAPTER 12

Pecking Orders in Chickens

No, pigeons are not the most common bird in the world. Chickens are, with a population of over 2-1/2 billion in the United States alone. Chickens are descended from wild jungle fowl, a type of pheasant found in Southeast Asia and India. They have been domesticated for more than 4,000 years. Oddly enough, most people know little about their behavior.

Chickens live in **flocks,** groups of birds that live, feed, and move together. Very large flocks are usually split into subflocks of around 50 birds. Each subflock has its own territory and is dominated by a single male.

The social structure of chicken flocks is based on a **pecking order** system of dominance. Chicken A can peck chicken B, who can peck chicken C, and so on. The highest ranking bird, chicken A, cannot be pecked by any of the lower ranking birds. In actuality, a flock of chickens has two separate pecking orders, one for the males and one for the females. The top-ranking male, however, is dominant over the top-ranking female. Typically, conflicts occur among chickens of the same sex, but not between sexes.

Pecking orders in chickens, although stable, may change suddenly as a lower ranking chicken challenges and defeats a higher ranking bird. Pecking orders also may not be obvious all of the time. Sometimes they are evident only during feeding, roosting, drinking, and competitive situations.

Chicken pecking orders need not be linear. **Linear dominance hierarchies** assume that if chicken A can peck chicken B, and chicken B can peck chicken C, then chicken A can peck chicken C. In chicken flocks, however, this may not always be the case. For example, chicken F may challenge chicken B and win. Chicken F is then dominant over chicken B. The loser, chicken B may still be dominant over chickens C, D, and E.

The winner, chicken F, may still be subordinate to C, D, and E. This ranking is called a **nonlinear dominance hierarchy.**

An interesting aspect of pecking orders is that individual chickens know their rank and appear to recognize the other chickens in the flock. Studies have shown that chickens recognize one another by the shape and color of their heads and **combs.** Hens whose identities were disguised when scientists painted their combs were attacked as strangers by their flockmates. When the paint was removed, these hens were again accepted by their flockmates, and they automatically assumed their previous dominance rank.

What is the function of a social structure based on a pecking order? Some suggest that the function of pecking orders is to reduce outright conflict in a flock. Others view it as survival of the fittest. The top-ranking hen typically stands at the center of the flock. The lower a hen's rank, the farther it stands from the center. In the event of predation, those hens in the center are the least likely to be caught.

How do chickens communicate? Chickens rely on both auditory and visual communication signals. They use over 20 separate calls, including calls associated with distress, contentment, alarm, and aggressive behavior. Roosters, for example, threaten another rooster with a moaning growl sound and advertise their territories by crowing. In addition to pecking, hens use various forms of visual communication to signal dominance or submission. For example, a hen raises its **hackles** (its neck feathers) to warn another hen not to come closer (Figure 12.1). Just before it attacks, a hen puffs out all of its feathers. In general, dominance is communicated by a raised tail. Submission is signaled by a lowered head and tail, sleeked down feathers, crouching, and turning aside.

Purpose

To observe how a pecking order system of social structure works in a flock of chickens.

Observational Setting

This project must be conducted on a farm or in a setting in which chickens are housed together in a group. A group of ten or so chickens would be the easiest to observe. With a larger group of chickens, it is difficult to observe all interactions, and it is difficult to identify individ-

Figure 12.1. Hen with raised hackles.

ual chickens. Be sure that you obtain the permission and cooperation of the owner.

Materials

- ☐ package of thin, multicolor pipe cleaners
- ☐ binoculars (optional)
- ☐ notebook
- ☐ pencil
- ☐ cotton balls
- ☐ brightly colored, nontoxic, water-based poster paint
- ☐ unscented baby wipes

Procedure

1. *Marking the chickens for recognition.* With the assistance of another individual, catch each chicken. Put a different-colored pipe cleaner band around the leg of each chicken (Figure 12.2). Be sure that you twist it a number of times and secure it carefully (chickens peck at almost anything, and you want the band to stay on for a few days). If you run out of colors, you can use two different color combinations twisted together. From this point on, you can refer to the

Figure 12.2. Identification band on hen's leg.

chickens by their color name (for example, red, green, blue). If the chickens are in a large pen and can range far away from you, you may need binoculars to check the color of a chicken's leg band.

2. *Identifying the pecking order.* Using a data sheet like the following, record each instance in which one chicken pecks another. Look at the leg bands and note which chicken delivered the peck and which chicken was the recipient. In the data sheet, the **aggressor,** or pecking chicken, is indicated on the y-axis and the recipient of the peck is indicated along the x-axis. For example, if the Red chicken pecks the Blue chicken, you would record a mark in the Red-Blue cell to indicate that Red pecked Blue.

   Continue to record each instance of pecking on the data sheet until you have collected enough data so that a pattern emerges. An ideal time to observe the chickens is during and after their daily feeding, because competition should intensify their social interactions.

   NOTE: Remember that there are two hierarchies, one male and one female. However, there may be so few roosters in the flock that you may want to focus exclusively on the hens.

Date: ____________
Time: ____________

		x—Recipient Chicken				
		Red	**Green**	**Blue**	**Black**	**Pink**
y—Pecking Chicken	**Red**			X		
	Green					
	Blue					
	Black					
	Pink					

3. *Altering the pecking order.* Once you have determined the structure of the pecking order system, you should be able to clearly identify the top-ranking, or alpha, hen and the second-ranking, or beta, hen. With the assistance of another individual, remove the beta hen from the flock. While someone else holds the hen, gently paint the comb of the beta hen using a cotton ball to apply the paint on both sides. Gently restrain the hen until the paint dries. Then quietly reintroduce it into the flock.

   Observe the behavior of the flock and record all instances of pecking for 1 hour. At the end of the hour, recapture the painted hen and gently remove the paint using a baby wipe. Be careful to remove all traces of the paint. Then reintroduce the hen back into the flock. Repeat your observations for another 1-hour period.

Questions

Flock Social Structure

1. Describe the social structure you observed. Can you identify the alpha, beta, and **omega** (bottom-ranking) birds? Was it a linear or nonlinear dominance hierarchy?
2. Was the alpha bird ever challenged?
3. Were other behaviors, besides pecking, displayed as a means of challenging another hen?
4. Did you see any submissive behaviors?
5. Did pecking behavior or challenges increase around feeding time?
6. Did the alpha bird have access to more food than the other birds? Were some birds excluded from feeding?
7. Did the alpha bird spend most of its time near the center of the flock? Were the lower ranking birds kept out of the center of the flock, particularly during feeding?

Disguising the Beta Bird

1. Did the pecking order change when the beta bird's comb was painted?
2. Did the beta bird retain its previous status? Was the beta bird treated as an intruder?
3. Was the beta bird the recipient of more pecks? Did the beta bird peck other birds more than before?

4. Did the status of other birds change?
5. Do you think the beta bird was recognized by the other birds as a member of their flock?

Restoring the Flock to Normal

1. Did the pecking order change when the paint was removed and the beta hen was returned to the flock?
2. Did the behavior of the beta bird change?
3. Did the behavior of birds that had previously challenged it change?
4. Based on your results, do you think that birds can recognize one another?

CHAPTER 13

Domestic Dogs and Their Masters

For over 10,000 years, humans have had a close relationship with domestic dogs. Domestic dogs were descended from wolves some 8,000 to 12,000 years ago. They have been selectively bred for all kinds of body types and temperaments. Domestic dogs perform numerous roles in human society as guide dogs, hunting dogs, rescue dogs, racing dogs, show dogs, police dogs, and shepherds, to name a few.

Dogs' wolf heritage is very much evident in their behavior. Wolves are very social animals that live in packs with a clear social organization (see Chapter 3). Wolves are competitive, yet little fighting is observed within their social group. They hunt cooperatively and jointly care for their young. Domestic dogs are also pack oriented. To domestic dogs, all humans are pack members, and their master is the leader of the pack.

The pack behavior of dogs is developed during a socialization period that lasts from about 4 to 12 weeks of age. From 4 to 6 weeks of age, their primary orientation is their littermates. They engage in play-fighting and establish a loose social organization based on dominance. The behaviors they learn affect how they interact with other dogs throughout adulthood.

By week 6, their interest in people and objects beyond their litter develops. For the next 6 weeks, their eyes and ears become sharper, and they show greater use of body postures, facial expressions, and vocalizations. If a puppy has individual contact with a human during this time, it will come to recognize that individual as its master and the leader of the pack. So important is their relationship with humans that puppies quickly learn to detect subtle changes in the eyes, face, body posture, and tone of a person's voice.

Dogs have a diverse language of their own. They communicate their social position, intentions, and emotional states through the use of sound, smell, and sight. Dogs can hear much higher frequencies than humans. They use auditory signals, communicating vocally through growls, barks, whines, and howls. The tone of a dog's bark expresses its meaning. High-pitched barks are used in greetings; deep barks are used as a warning; and still deeper barks, often alternating with growls, are used aggressively as threats.

The **olfactory sense** of dogs is much more highly developed than in humans. Dogs have a second smell organ in the roof of their mouth and use an anal scent gland for scent marking. From a drop of urine or a small amount of feces, another dog can tell the dog's sex, whether it was castrated or spayed, how recently the scent mark was left, the direction the dog was going, and, if it is a female, whether or not the dog is in season.

Dogs have evolved an extensive system of visual communication using facial expression, ear position, tail position, and body posture. Every time a dog comes into contact with a "pack member," be it human or another dog, it uses its entire body to greet the pack member. The following are some commonly used greetings.

Aggressive threat. This behavior is shown by the pack leader or by a dominant dog when approaching another individual (Figure 13.1A). The dog's ears are erect and pointed forward, with the mouth slightly open and the lips curled to reveal the canine teeth. The facial muscles on the nose and forehead are wrinkled, and the eyelids are raised. The tail is typically erect and wags in tight circles. The neck and body are erect, the legs are stiffened, and its hackles (body hair on its neck and back) are raised.

Active submission. This behavior is shown by a dog to signal that it wants to approach a more dominant individual. The dog's ears are pointed backward, and the mouth is open in a long grin. Sometimes, it may attempt to lick the face of the other individual. Its body is crouched low to the ground (Figure 13.1B). Its tail may be tucked under, between its legs, or the tail may be wagging anxiously.

Passive submission. This behavior is shown by a dog that has no other choice than to show submissive behavior (or else be attacked). This behavior occurs typically when the dog is approached by a dominant, threatening individual. The dog's ears are flattened against the side of its head, and eye contact with the more dominant dog is averted. The dog drops to the ground, rolls on its side, and reveals its genitals and stomach. Its front legs are typically bent and held

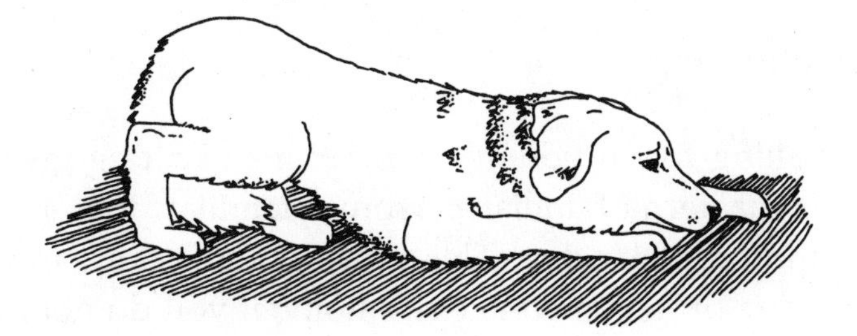

Figure 13.1. Aggressive threat, active submission, and play-initiation postures of dogs.

limply in the air, and its tail is tucked very tightly under. It may also urinate.

Play initiation. This behavior is frequently shown by puppies but may also be observed in older dogs (Figure 13.1C). The dog raises its ears and opens its mouth in a long, relaxed grin without showing teeth. It lowers the front half of its body and nearly touches the ground with its chest. The dog wags its tail and sometimes its entire hindquarters, excitedly. It may nudge or paw the individual or present it with an object like a stick or toy.

Purpose

To observe how a domestic dog greets different types of individuals.

Observational Setting

You will be making five different observations of a dog greeting other individuals: its master, a familiar person, a familiar dog, an unfamiliar person, and an unfamiliar dog. This project can be done in your home with your pet dog or in someone else's home if you do not have a dog. You may use either an adult dog or a puppy (four months or older).

NOTE: All of the observations should be made in the dog's home territory to control for unfamiliar settings and odors. Also, be sure that you only use dogs known to be friendly as test subjects.

Materials

☐ an adult dog or a puppy at least four months old

☐ notebook

☐ pencil

Procedure

1. *Observing facial expression and body posture.* For each of the observational situations, describe the following aspects of the dog's facial expression and body posture.

Ears. Are the ears erect? Do they point forward or backward?

Mouth. Is the mouth open? Are the lips curled? Are the muscles around the mouth tense or relaxed? Does the dog lick the individual?

Teeth. Do the teeth show? Are the canine teeth exposed? Are the upper and lower teeth together or apart?

Head position. Is the head held forward or backward? Is it lowered or held high?

Tail position. Is the tail wagging or still? If wagging, are the circles loose and wide or are they small and quick? Is the tail pointed? Is it erect? Is the tail held low or high, or is it tucked between the dog's legs?

Hackles position. Are the hairs on the dog's neck, back, and rump erect?

Legs position. Are the dog's legs bent or straight? Are they stiff or limp?

Body position. Is the dog's body partially or fully lowered? Is the dog standing or on its back? Is the dog moving its body, or is it still? Does the dog rub any part of its body against the individual?

2. *Observing greetings with persons.* Start your first observation with an *unfamiliar* person in the dog's home territory. Lead the dog to a position approximately 4 feet (1 m) from the individual. Try to use someone of the same age and sex as the dog's master. Instruct the person to crouch down slightly, make eye contact with the dog, and extend his or her arms toward the dog (Figure 13.2). Ask the person not to use any verbal or auditory commands. Observe how the dog initially responds to the person. Then observe how it responds 5 minutes later. Describe how the person interacts with the dog (for example, does the person pat the dog? stand back from it?).

   On a different day, in the same location, repeat the preceding observation with (a) a person *familiar* to the dog and (b) the dog's master. (If you are the dog's master, work with a friend to make observations for you. Try to make sure that person records observations consistent with your way.)

3. *Observing greetings with other dogs.* Start your first observation with an *unfamiliar* dog in your dog's home territory. Have someone lead the unfamiliar dog to a position approximately 4 feet (1 m) from your dog. Try to use an unfamiliar dog of approximately the same

Figure 13.2. Dog greeting an unfamiliar person.

size, sex, and age as your dog. Release the two dogs at the same time and observe how your dog initially responds to the unfamiliar dog. Then observe how it responds 5 minutes later. Briefly describe how the unfamiliar dog interacts with your dog.

On a different day, in the same location, repeat the preceding observation with a *familiar* dog.

4. *Summarizing the data*. Summarize the data you collected. Try to characterize the type of greeting you observed in each situation as an aggressive threat, active submission, passive submission, or play initiation.

Questions

Type of Response

1. Did the dog respond in a similar way to both the dogs and people?
2. Did it respond to its master differently than it did to other dogs and people? Why might you expect it to have a unique relationship with its master?
3. In general, how does the dog respond to familiar versus unfamiliar individuals? Why might it have a characteristic way of approaching unfamiliar individuals?
4. Do you think that an unfamiliar dog would be more of a threat than an unfamiliar person? Why?

5. Would you be more likely to see passive submission with an unfamiliar dog or an unfamiliar person?
6. Why do you think that two forms of submissive greetings (active and passive) have evolved?
7. Why doesn't the more dominant dog attack when approached in a submissive posture?
8. Why do you think puppies are more likely to show passive submission?
9. Did you observe any instances in which submissive behavior shifted to aggressive threats or a different type of greeting over time?
10. What happened 5 minutes after either dog exhibited aggressive threat behavior?

Tail Position

1. One of the easiest ways to instantly tell the relationship between two dogs (or a dog and a human) is to focus on the position and movement of the dog's tail. Did you find the tail to be an accurate indicator of the relationship?
2. Why do you think that dominant dogs erect their tails while more submissive dogs hold them closely to their bodies?
3. Given that dogs have an anal scent gland, do you think the tail is used for olfactory communication as well as visual communication?

CHAPTER 14

Pigeon Courtship

Millions of people work in cities every day and share sidewalks, parks, and office buildings with pigeons. Pigeons are the most obvious form of wildlife in cities, but few people have taken the time to observe and understand their behavior.

Pigeons are the first known bird to be domesticated by humans, perhaps as early as 4500 B.C. Originally, they lived on cliffs and rocky ledges overlooking the sea. There, they developed the ability to navigate and to land on narrow cliffs in strong winds. They are one of the strongest and fastest flying birds, achieving speeds of over 80 miles (129 km) per hour.

Pigeons are best known for their navigation instincts. Long before telephones, radio waves, and air travel, they were used to carry long-distance messages. Scientists have shown that pigeons navigate using the sun as a directional guide. In recent years, scientists have hypothesized that they navigate by smell as well, forming a map of odors by learning scents associated with each compass orientation.

The next time you see a flock of pigeons swoop down on the sidewalk in front of you, look more closely. Are they just moving about randomly, scouring the streets for food, or is there some pattern to their behavior?

Most of the behavior you see in a flock of feeding pigeons is probably courtship-related. In the early stages of courtship, their behavior may seem aggressive. As courtship progresses, it becomes more and more sexually oriented. Unfortunately, the majority of courtship behavior occurs in the nest or at the nest site. Occasionally, however you might observe **mounting** behavior, or attempts at mating. Pigeons in cities roost in such places as window ledges of 30-story buildings and on the spans of suspension bridges. Thus, this project will observe the beginning of courtship behavior in a flock setting.

A classic study of pigeon courtship behavior by Fabricius and Jansson outlines a number of behaviors shown during the early stages of courtship. Some of these are summarized in the following list. Because male and female pigeons look alike, it is through these behaviors that males and females may be identified.

Bowing. The bird raises its neck feathers while bowing its head and points its bill downward (Figure 14.1A). Then, it rises, spreads its tail, and drags it on the ground (Figure 14.1B). As it rises, it stretches its neck up so that its body is almost vertical. *Bowing is mostly a male behavior. The tail-drag component is exclusively a male behavior.*

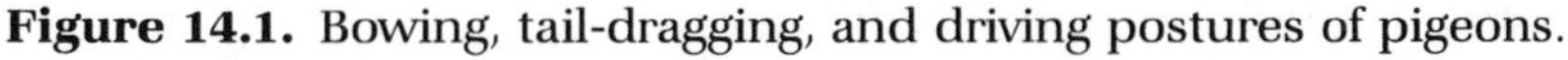

Figure 14.1. Bowing, tail-dragging, and driving postures of pigeons.

Figure 14.2. Pigeon demonstrating wing clapping.

Attacking intention. The bird raises its feathers, pecks in the air in front of another pigeon, and vibrates its wing on the side closest to the opponent.

Driving. The male raises its neck, points its bill down, and runs closely after the female, often treading on her tail (Figure 14.1C). *Driving is exclusively a male behavior.*

Begging and billing. The female pecks at the feathers surrounding the base of the male's bill. Then, the male opens its beak and the female inserts its bill as both birds perform pumping and twisting movements. The female sucks food from the crop of the male. *Begging is exclusively a female behavior.* These behaviors may be observed during courtship, but typically they are nest-related.

Strutting. The pigeon raises its neck and walks back and forth around the female.

Wing clapping. The pigeon makes deep wing beats as it takes off to fly. The wing beats make a clapping sound (Figure 14.2).

Purpose

To observe pigeon courtship behavior in a flock and to identify males and females on the basis of behavior.

Observational Setting

The pigeon breeding cycle typically starts in spring and extends through summer. In warmer climates, courtship behavior may be seen year-round. It is probably easiest to start your observations in spring. Choose any safe location in a city. It might be near a park bench, on the side-walk, next to a ballfield, or any place that pigeons may gather. A great place to observe is where people scatter bread crumbs for them. Make sure you are not sitting under a place where they roost, for that can be a messy experience!

Materials

- ☐ watch
- ☐ bread crusts or crumbs
- ☐ notebook
- ☐ pencil

Procedure

1. *Observing the flock.* Find a flock of pigeons to observe. Choose 1 hour during midmorning. Spend an hour just watching the birds and becoming familiar with their behavior. Then, observe the flock of pigeons at the same time daily for five days.

   NOTE: The group of pigeons may change daily, but you should return to the same site. Bring some bread crusts to attract a flock of pigeons, if necessary.

2. *Observing and recording courtship initiation.* Observe the flock until you see a pigeon initiating one of the following behaviors:
   - bowing.
   - attacking intention.
   - driving.
   - begging.
   - billing.
   - strutting.
   - wing clapping.

Frequency of Pigeon Behavior in a 1-Hour Period

Date: ______________

Time: ______________

	Recipient Reaction							
Behavior of Initiator	**Strutting**	**Begging**	**Billing**	**Pecking**	**Wing Clapping**	**Fly Away**	**Ignore**	**Other**
Bowing								
Attacking Intention								
Driving								
Begging								
Billing								
Strutting								
Wing Clapping								

If that pigeon has distinctive coloring, markings, or some characteristic to help identify it at a later time, make a note of it.

3. *Observing and recording recipient reaction.* Follow the activities of the initiator and determine which pigeon is the recipient of the behavior. Again, note any distinctive features, if possible. Record the reaction of the recipient bird using a data sheet similar to the one on page 100. Place a check mark in the appropriate box each time a behavior occurs in each 1-hour observation period.

   NOTE: Try to follow one pair of birds until one or both flies away.

Questions

Observing Behavior

1. Look over your data sheets, comparing the responses to each behavior. Based on your results, what do you think each of the behaviors means?
2. Do you think that some of the behaviors are communicating more than one meaning? Why might they have more than one meaning? For example, research on courtship behavior in a number of animal species suggests that there is a fine line between sex and aggression. A male pigeon may bow to a female to initiate courtship behavior, or he may bow to a male pigeon to intimidate it.

Identifying Aggressive and Courtship Behaviors

1. Try to categorize those behaviors that elicited a negative response as aggressive. What was the most frequent negative response? How often were aggressive behaviors simply ignored?
2. Categorize those behaviors that elicited a positive response as courtship-related. What was the most frequent positive response? Did you see any overtly sexual behavior, such as **mounting**? What happened when courtship behavior was ignored? Did the initiator persist or direct its attention toward another bird?
3. If you could identify any birds individually, did they always initiate interactions? Did they show the same behaviors?

Identifying Males and Females

1. Based on the following assumptions of gender and behavior, could you identify the differences between males and females?

Male behaviors. Tail-drag, driving.

Likely male behaviors. Bowing, strutting, wing clapping.

Female behaviors. Begging.

2. Do you think males can only be distinguished from females during courtship?

CHAPTER 15

Nonverbal Communication in Human Interactions

It has been said that we speak with our vocal cords, but we converse with our whole body. A typical person speaks, on average, for only 10 to 11 minutes per day. Each standard spoken sentence lasts for approximately $2^1/_2$ seconds. Why does it seem that we spend far more time talking? Most likely, it is because so much of our communication is nonverbal. Nonverbal communication is the use of our body, our demeanor, and our facial expressions to express ourselves. It is estimated that in a normal, two-person dialogue, the conversation is made up of 35% verbal components and 65% nonverbal components.

The function of verbal communication is to convey information. Nonverbal behavior conveys information as well, but it is used primarily to manage the relationship aspects of the communication. The principal kinds of nonverbal communications used in human interactions include the following:

Appearance. The effect of an individual's appearance sends messages either *voluntarily* (for example, through things we can manipulate, such as clothes style or hair cut) or *involuntarily* (through things we do not control, such as height and physique).

Proximity. The distance maintained between individuals during interactions.

Orientation. The angle at which people sit or stand during interactions, ranging from head-on to side by side.

Bodily contact. The degree to which individuals make contact, ranging from light brushing to hitting and pushing.

Posture. The various ways that individuals can stand, sit, or recline.

Gestures. The way individuals use their head, hands, feet, and other body parts to indicate meaning. They can be either deliberate movements or random actions.

Facial expression. The combined use of the eyes and facial muscles to communicate attitudes and emotions.

Nonverbal components of speech. The variations of pitch, stress, timing, tone of voice, and accent in verbal communications.

Human speech appears to be coordinated with body movements. These body movements, including the use of the head, hands, and eyes, synchronize the actions of the speaker and listener. Typically we are not aware of how we use body movements. In addition, they are usually only visible in our peripheral vision as we converse.

When strangers meet, it is unlikely that their natural speaking styles match. Before the conversation flows smoothly, there is an adjustment period in which one person speaks faster, listens more, and so on. As the conversation progresses, nonverbal communication is used to signal whose turn it is to speak. More specifically, certain body movements, known as **interaction markers,** assist us in taking turns (Figure 15.1). Common interaction markers are shown in the following table.

Common Nonverbal Interaction Markers

Verbal Behavior	Nonverbal Activity
End of statement	Downward head movement
	Downward eyelid movement
	Downward hand movement
End of question	Upward head movement
	Upward eyelid movement
	Upward hand movement
End of argument	Head turn or tilt
	Neck flexion or extension
Shift point of view	Lean back to listen
	Lean forward to talk

Nonverbal communication also is used to provide feedback during a conversation. During conversations, we unconsciously use facial expressions to provide the other participant with feedback. Our expressions convey understanding, disagreement, interest, pleasure, and so on. (When we speak on the phone without the normal visual cues, we find

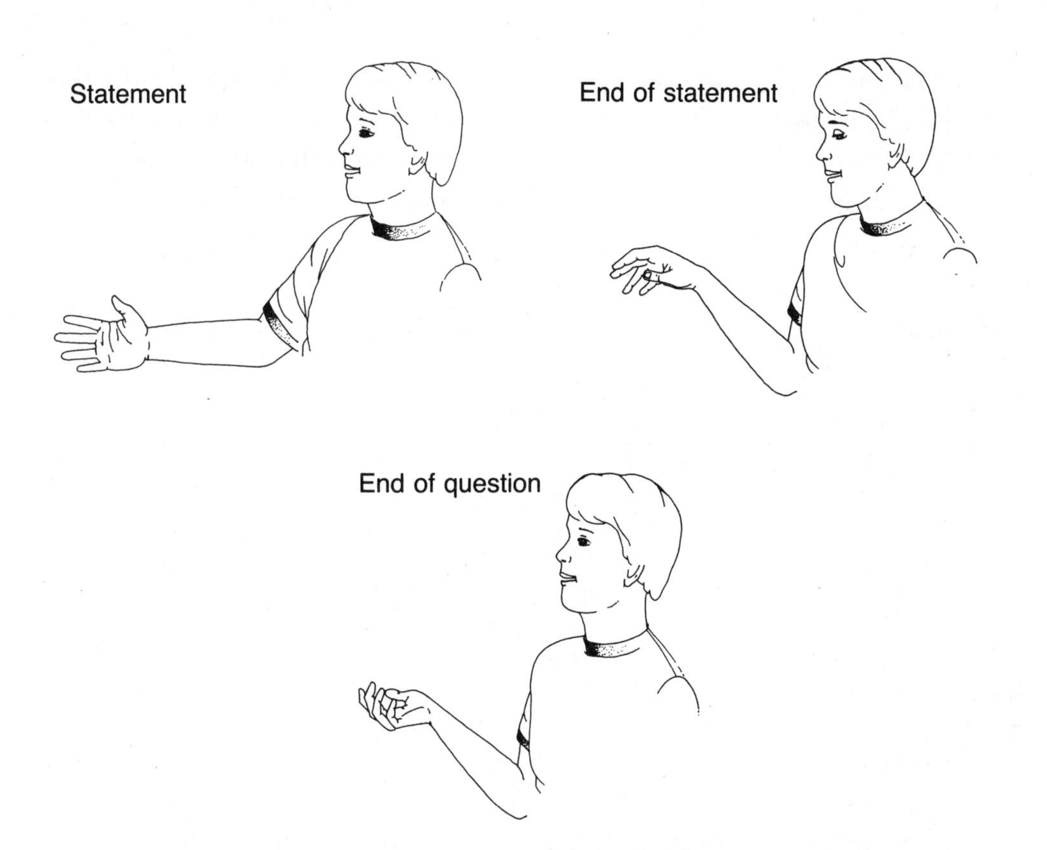

Figure 15.1. The use of common nonverbal interaction markers.

ourselves using verbal cues such as "Really?" "Umhum," and "okay?" for feedback.)

Nonverbal communication in humans is a vast area for study. With over 180 physically distinct forms of smiling alone, imagine how many nonverbal behaviors we could categorize. Thus, this project will focus on a small aspect of nonverbal communication—how we use interaction markers in conversations.

Purpose

To observe how interaction markers are used in conversations between familiar and unfamiliar individuals.

Observational Setting

This project is best done by observing videotapes of conversations. By using a videotape, you can stop the action and replay sequences to get a very accurate view of the nonverbal behavior. You also may do this project by watching live conversations; however, subtle behaviors will be difficult to observe, and your presence may affect the behavior of your subjects. For the purposes of this project, it is assumed that a VCR is used.

Materials

- ☐ VCR
- ☐ VCR tapes
- ☐ television
- ☐ notebook
- ☐ pencil
- ☐ stopwatch, or watch with second hand

Procedure

1. *Choosing the conversation tapes.* You may use any source for the conversations. You might select scenes from a movie, sequences from talk shows, scenes from home movies, videotapes of your friends or schoolmates, or excerpts from the news. Bear in mind that you may be observing "professional conversationalists," such as talk show hosts. Such individuals are often very aware of nonverbal communication cues and use them for dramatic effect.

   View the tapes until you find a conversation for each of the six categories shown in the following table:

Conversation Type	Sex of Participants		
Familiar individuals	M–M	M–F	F–F
Strangers	M–M	M–F	F–F

   Try to control for similar contexts by following a few rules:

   - The conversation should involve only two individuals.
   - The conversation should last at least 3 minutes.
   - The conversation should take place indoors.

- Both participants should be in the same position (for example, sitting, standing).
- The source (for example, movie, talk show, or news) should be the same for all conversations, if possible.

2. *Describing the context.* To interpret your data properly, it is important to note as many details about the context as possible. Some examples are the following:

The setting. Are there other people around the participants who might influence their behavior? How far apart are the participants? Is there furniture or space considerations that may affect behavior? Are one or both of the participants in a familiar setting?

The participants. Approximately what age are the participants? Will status or authority affect their relationship? Describe their appearance—is their hair style, dress, and so on likely to affect their behavior? Are they from similar socioeconomic and cultural backgrounds? If they are familiar, what is the relationship of the two participants?

3. *Observing and recording the data.* For each conversation, label the individuals as A and B (or male and female, if appropriate). Prepare a separate data sheet for each participant. You always will be watching and recording the behavior of the *talker*. Record interaction markers displayed by the talker whenever he or she does the following:
   - ends a statement.
   - ends a question.
   - ends an argument (or series of statements).
   - switches roles (for example, from talker to listener).

   Record the following interaction markers:

   Body orientation. Forward (F) or backward (B).
   Head position. Down (D), up (U), or level (L).
   Eyelids position. Down (D), up (U), or level (L).
   Hands position. Down (D), up (U), or level (L).

   A sample data sheet is shown on the next page to help organize your data. You would have a separate one for Subject B.

Subject: A (Female)
Conversation Type: Familiar, female–female
Context: Females are both in their mid-thirties. They are dressed in casual clothes. They are seated across from each other in chairs and appear to be friends, and so forth.

Verbal Event	Body Position	Head Position	Eyelids Position	Hands Position
Statement	F	D	D	D
Statement	F	D	L	D
Question	F	U	U	D
Argument	F	L	L	D

Notes:

Questions

Use of Interaction Markers

1. Review all of your data sheets, regardless of the sex or familiarity of the participants. Were interaction markers used more often than not?
2. Were the markers more obvious for certain verbal statements (such as questions) than others?
3. Were the markers you observed consistent with your predictions?
4. Did the subjects appear to spend more time communicating verbally or nonverbally?
5. Did you observe other nonverbal behaviors that made the conversations flow more smoothly?

Behavior of Familiar versus Unfamiliar Individuals

1. Which group would you expect to make greater use of interaction markers? Separate your data sheets into two categories: familiar participants and unfamiliar participants. Did both groups use interaction markers? Did one group use the markers more frequently than the other?

2. Did their use of nonverbal cues change over time? For which group would you have expected to see a change in the use of interaction markers over time?
3. Review the contexts for each of the conversations. Can you relate your results to the contexts?

Behavior of Males versus Females

1. Separate your data sheets into males and females. Do you see any pattern in the way males or females use interaction markers? In general, do you think that males or females rely more on nonverbal communication? Did your results support your belief?
2. Does the use of interaction markers by males and females change depending on the familiarity of the individuals?

PART IV

Eating and Sleeping: Projects on Feeding and Nesting Behavior

One of the fundamental decisions an animal has to make is where to live. The fact that some species are found only in particular habitats suggests that animals actively select their nesting site, while avoiding or ignoring others. An animal spends most of its life within a geographically defined area. This area is its home range, which supplies its basic requirements for food and shelter.

Home ranges vary in size, depending on the availability of food and water and the time of year. Many animals during mating season actively defend their home range or nesting site. This behavior, known as **territoriality,** is typically associated with competition for food sources, breeding sites, or breeding partners. Siamese fighting fish, for example, are territorial in their native habitat. In the presence of a receptive female, the male builds a bubble nest. After the female ejects her eggs, the male places them in the nest and actively defends the nest site against any intruder, male or female, until the young hatch.

The most common activity of many animals is eating. Some animals must eat almost continuously in order to survive. Others are active and eat at certain times of day. For the most part, animals develop a feeding strategy that maximizes the amount of food they obtain with a minimal amount of energy expenditure. Bumblebees, for example, have high energy requirements. They cannot afford to expend all of their energy searching for and collecting food, so they have developed feeding strategies based on the color and shape of a flower.

Feeding behavior is not without its risks. Animals often come out into the open to search for food and are more visible to **predators** (animals that capture and feed upon them). Gray squirrels, for example, have a number of natural predators, regardless of whether they live in the woods or in a city park. They must always balance their need for food with the risk of predation. As a result, they alter their feeding strategies depending on their distance from protective cover.

Some animals, such as domestic cats, are **carnivores** and hunt their prey. Others, such as horses and sheep, are **herbivores,** and spend much of their day grazing and digesting food. Still others, like spiders, have developed amazing mechanisms for capturing prey. Some spiders construct webs that allow them to detect their prey through vibration of the web. Other spiders respond to changes in the tension of their web resulting from broken threads.

Why do animals choose certain nesting sites? How do they move about their home ranges? Do they use strategies based on the availability of food and the risk of becoming someone else's food? These are some of the questions you might ask in observing the eating and nesting patterns of animals.

CHAPTER 16

Spider Webs at Work

There are over 30,000 known species of spiders. For centuries, they have fascinated individuals with their ability to spin elaborate silk web designs. The average spider web contains about 75 feet (23 m) of silk and consists of over 1,000 junctions. Webs range in complexity from a simple silk line to complex, three-dimensional webs.

The primary purpose of spider webs is to capture food. The type of web a spider spins is a function of the spider's physiology, characteristics of its food source, and the type of environment in which it lives.

Spider silk is a remarkably strong fiber, with a **tensile strength** (resistance to stress) far greater than that of steel. For example, the web of a large tropical spider, *Nephila,* is strong enough to capture small birds. It has been used by primitive peoples to make fishing nets, lures, and tightly woven bags for carrying such things as arrowheads and dried poisons.

Spider webs can be classified into four basic types: **funnel webs, sheet webs, orb webs,** and **irregular mesh webs.** Funnel webs (Figure 16.1A) are platformlike webs with a tube or funnel used by the spider as an exit. Funnel webs may be located just above the soil or suspended in the air like a hammock. Typically, insects mistake the web for a landing area. Once they land, they sink into the spongy trap. The spider runs over, captures them, and drags them into the funnel.

Sheet webs (Figure 16.1B) also contain a platform. The spider clings upside down to the bottom of the web and waits for its prey. Typically, insects drop into the sheet after being knocked down by a mass of crisscrossed lines suspended above the web. When an insect falls, the spider attacks its prey and pulls it through the webbing. Sometimes a sheet web contains a second sheet beneath the spider, which serves as a form of defense from below.

Orb webs (Figure 16.1C) are the traditional spider webs of Hal-

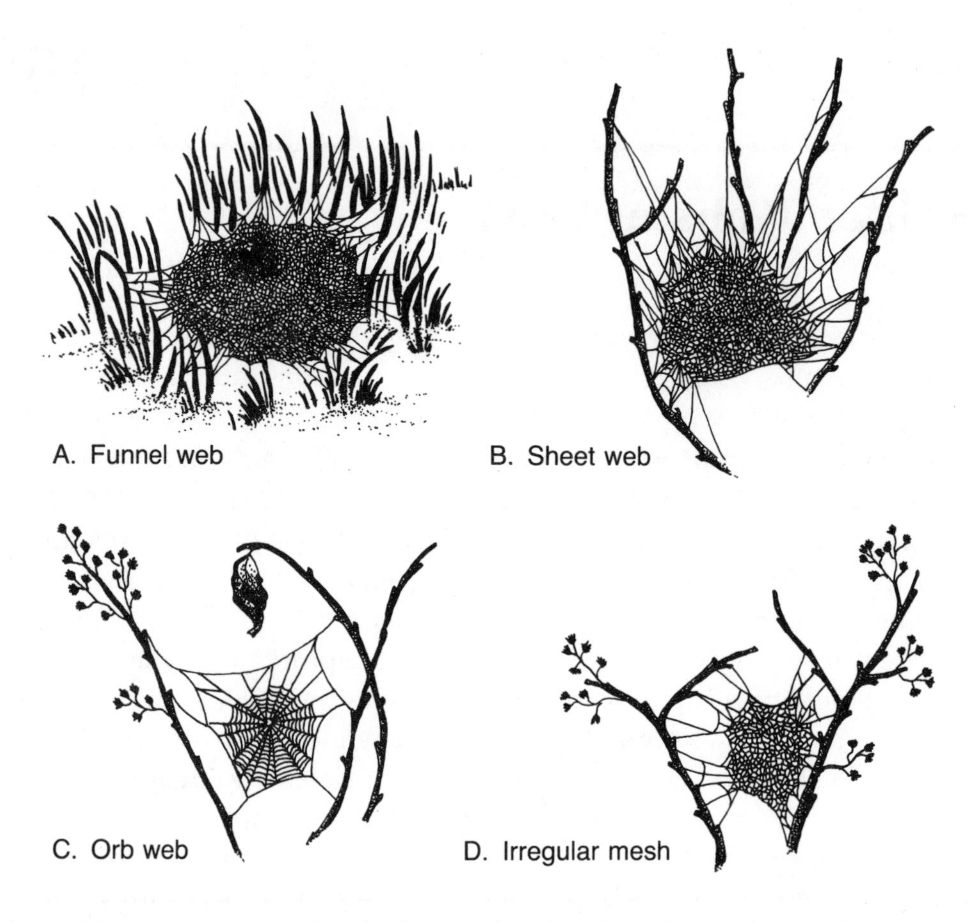

Figure 16.1. Funnel web, sheet web, orb web, and irregular mesh web.

loween and *Charlotte's Web* fame. They are anchored by a number of **radial lines,** or **radii,** of dry silk. At the center of the web is a meshed **hub.** A series of sticky **spiral coils,** or **capture threads,** are attached to the radii and used for trapping insects. The spider typically hangs in the center of the hub with its claws touching the radii. It is often attached to a **drag line,** a single strand of silk used to drop down from the web. When an insect gets stuck in the web, the whole web vibrates. The vibrations are communicated to the spider, who rapidly moves to the site of the disturbance. The spider then uses its front legs to capture the victim. Then, it turns it round and round, wrapping it with silk.

Irregular mesh webs (Figure 16.1D) have no particular design. Spiders typically hang upside down beneath the mass of silk to wait for insects to get caught.

In all the web examples, the presence of an insect is communicated to the spider in one of two ways. The spider either feels vibrations (as in the orb web) or it senses changes in the tension of the web resulting from broken threads (as in the three-dimensional designs). Dislodging or breaking strands is typically what stops the insect in three-dimensional webs. In orb webs, however, insects could simply fly through broken threads and escape. Instead, scientists have discovered that the silk of orb-weaving spiders has incredible elasticity. Some threads can be stretched to over four times their original length without breaking.

What makes the silk of orb-weaving spiders so elastic? Scientists have found that some garden spiders have special glands that apply a liquid to the surface of the spiral coils. This liquid attracts water in the air. The water quickly separates into little droplets. This is typically how we are able to spot so many spider webs on a dewy morning. Although silk is quite stiff when dry, it is much more elastic when wet. Thus, it can absorb far more of the energy created by a struggling insect.

Purpose

To observe how spiders use their webs to capture insects.

Observational Setting

Spider webs can be found practically anywhere—indoors, outdoors, in barns or garages, in grassy areas, in the woods, or anchored in bushes during warm months.

Materials

- ☐ notebook
- ☐ pencil
- ☐ fine watercolor brush
- ☐ spider identification guide (optional)

Procedure

1. *Locating different webs.* Take a walking tour and try to locate as many examples as you can of a funnel web, a sheet web, an orb web, and an irregular mesh web. Try to locate the spider. Note the following:
   - the type of web (you might try to sketch it).
   - the location of the web.

- the location of the spider in the web.
- the physical characteristics of the spider (coloring, body type, and so on)
- the presence of insects or other objects in the web.

2. *Locating orb webs.* Take a walking tour early in the morning or after a rain shower. Try to locate as many examples of orb webs as you can. Note the following:

- Are there water droplets on the radii?
- Are there water droplets in the hub?
- Are there water droplets on the spiral coils or capture threads?

3. *Testing the spider's reactions.* Find an active orb web and locate the spider. It may be necessary to stimulate the web to bring the spider out of hiding. Test the conditions shown in the following list. Try not to scare the spider. Move slowly, and avoid casting a shadow over the web. Wait several hours or up to a day between the conditions. If necessary, use different spider webs.

a. Blow on the web.

b. Using the brush, gently vibrate one of the radii; then vibrate some capture threads (Figure 16.2).

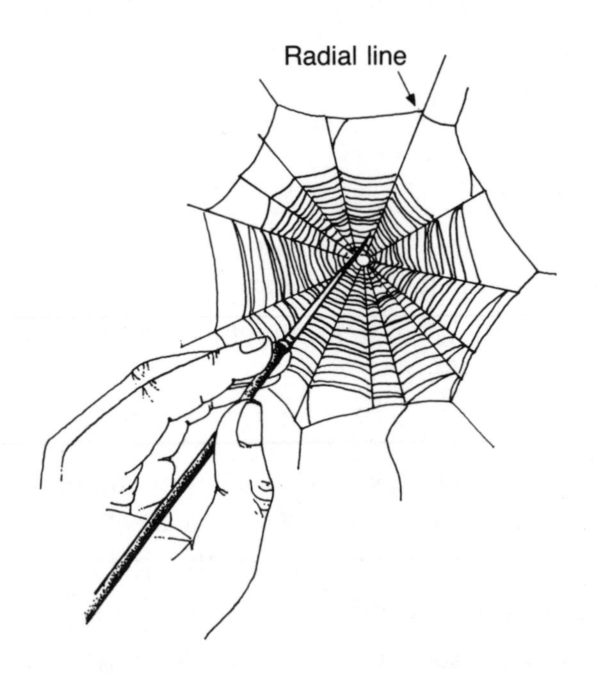

Figure 16.2. Vibrating the radii with an art brush.

c. Gently flick a dead fruit fly or some other small insect into the web. (Look for small insects in the center of wildflowers.)

d. Remove a portion of a wing of a live fruit fly or some other small insect so that it cannot fly. Gently flick this insect into the web.

e. Gently try to break a few strands of the capture thread.

Questions

Spider Web Characteristics

1. Were you able to locate the spider in the webs you observed? You might have seen it in the funnel of the funnel web, clinging to the bottom of the sheet web, or in the center hub of the orb web.
2. If you did not see the spider in the orb web, did you see any evidence of a drag line? An orb weaving spider may hide off to the side or below the web, connected to the web by its drag line.
3. Did you observe other insects or inanimate objects in the web? Were the other insects wound in silk? Would you expect to find inanimate objects in the web? Why?
4. Did you see more droplets of water on the capture thread of the orb web? Assuming it was a garden spider's web, you probably did because the coating on the capture threads actually attracts atmospheric water. The capture threads are soft and highly elastic. But once the insect sticks to the web, it is the water droplets that help retract the capture threads back into place.

Spider Reactions

1. When you blew on the web, how did the spider respond? Did it react at all? Did it respond defensively by moving to the side of the web? Why might the spider respond only slightly to the overall movement of its web?
2. When you tapped the web's radii, did the spider react? Did it move to the center of the web? Did it move toward the source of stimulation or away? How do you think the spider would respond if you repeatedly tapped the radii?
3. Did the spider react differently when you tapped the capture threads? Did it move toward the source of stimulation, or did it move toward the center of the web? Why might the spider respond differently to stimulation of the capture threads and the radii?

4. Compare how the spider responded to the dead fruit fly and the live fruit fly. Why might it have responded differently? Even though a dead fruit fly is a source of food, why might the spider ignore its presence in the web? Do you think the dead fruit fly would still be in the web 24 hours later?
5. What happened when you broke a few strands of the capture threads? Did the spider react by moving to another area of the web or by leaving the web all together? Why might an orb-weaving spider not respond to a tear in the capture threads? Would you expect a spider that spins a three-dimensional web to respond to a tear in its web?

CHAPTER 17

Feeding Strategies of Gray Squirrels

There are over 1,000 kinds of squirrels in the world today. The gray squirrel is the most widely distributed species of tree squirrels in North America.

The gray squirrel has gray fur with buff-colored underfur and a bushy tail. It uses its tail for balance in jumping, climbing, running, and turning quickly. If a squirrel falls, its tail acts as a parachute and slows its descent. Gray squirrels are active during the day. During most of the year, they have two peak periods of activity: for several hours after sunrise and in midafternoon. During winter, gray squirrels are active around midday.

Individual squirrels have home ranges that vary in size according to the abundance of food, the availability of cover, and the season of the year. In general, home ranges are several acres and are larger for males than for females. The gray squirrel's social group is made up of all the squirrels that overlap its home range. Some evidence indicates that individual squirrels recognize each other through sight, sound, and smell.

Gray squirrels live and nest in wooded areas They sleep and rear their young in nests made of clumps of leaves and twigs or in tree-hole dens lined with leaves. Gray squirrels do not store food in their nests and must go to the ground to hunt for food.

The diet of the gray squirrel varies throughout the year. Their main source of food is nuts. Throughout fall and winter, they eat all different kinds of nuts, including acorns, hickory nuts, walnuts, and beechnuts. In a one-week period, the typical adult may eat its own weight in nuts. Squirrels tend to maximize their **feeding efficiency,** expending the least amount of energy to obtain the maximum amount of food.

Gray squirrels gather nuts and bury them one-by-one in the ground. They dig a hole 3 to 4 inches (8 to 10 cm) deep and push the nut down into the ground with their nose. So efficient is their burying ability that they can locate and bury a nut in 60 seconds. It is thought that they scent-mark the nut with their nose as they bury it so they can dig it up at a later time. In winter, they can smell the location of buried nuts through a foot of snow.

In fall, gray squirrels also eat apples, pinecones, and sometimes conifer bark. Their spring diet consists of buds of oak and hickory trees; the inner bark of some hardwood trees; ripened seeds of elm, maple, and basswood trees; and sometimes songbird eggs. In summer, they eat flowers, early fruits, mushrooms, beetles, and caterpillars. If they live in the country, they also may eat field corn. On the other hand, if they inhabit the city parks of major metropolitan areas, they probably eat popcorn, peanuts, pretzels, bread crusts, and diet soda!

Squirrels have a number of natural predators. These include hawks, owls, foxes, coyotes, snakes, skunks, bobcats, raccoons, badgers, and domestic dogs and cats. Whether they live in city parks, suburban locations, or the country, squirrels must always balance the need to move about, searching for food, with the risk of predation. The availability of food, as well as the pressure from predators, differs in each of these habitats. As a result, the number of squirrels in a particular area, the size of their home ranges, and the ease with which they meet their food requirements vary, as well.

Gray squirrels give alarm calls (churring and tail waving), but they do not depend on their social group for defense against predation. They can see in almost every direction without moving their heads, and they are constantly aware of movement in their environment. The squirrel dashes to the nearest tree when threatened. It flattens its body against the opposite side of the trunk of the tree and freezes until the danger is past. Gray squirrels also use "highways." They habitually follow the same route again and again, jumping from tree to tree. These highways appear to be learned and allow them to move rapidly under cover from one area to another.

Studies show that a squirrel's behavior while searching for food is highly dependent on its distance from cover. How does a squirrel balance the competing need to maximize its feeding behavior while avoiding predation? This project is designed to answer this question by varying the availability of a squirrel's food and the distance of the food source from cover.

Purpose

To observe how the feeding strategy of gray squirrels differs as the distance from cover increases.

Observational Setting

This project may be done wherever you can locate a population of gray squirrels. It may be done in a city park, in the suburbs, or in the country. In the city, block-long parks or squares make ideal settings because the squirrel density is usually very high. In any setting you choose, you need to find a wide-open area (such as a field or terrace) that is at least 40 to 50 feet (12 to 15 m) from the nearest tree or source of cover. You will be constructing two feeding "platforms," one with four feeding places (high availability) and one with one place to feed (low availability). These platforms will be placed at various distances from protective cover.

Materials

- ☐ five 9-inch (23-cm) aluminum foil pie plates
- ☐ two 3/4-inch-thick, 2-by-2-foot (2-cm-thick, 0.5-by-0.5-m) pieces of plywood
- ☐ hammer
- ☐ fifteen 1/2-inch (1-cm) nails with large heads
- ☐ 10-pound (4.5-kg) bag of unhulled sunflower seeds
- ☐ measuring tape
- ☐ binoculars
- ☐ notebook
- ☐ pencil
- ☐ stopwatch, or watch with second hand

Procedure

1. *Constructing the feeding platforms.* Construct two feeding platforms, one for a high food availability condition (four pie plates) and one for a low food availability condition (one pie plate). For the high food availability condition, attach a pie plate to each corner of a piece of plywood (Figure 17.1A). Use three nails to secure each pie plate. For the low food availability condition, attach a single pie plate to the center of the other piece of plywood, again using three nails (Figure 17.1B). Fill each pie plate with approximately 1 inch (2.5 cm) of sunflower seeds (or enough so that you do not have to refill the plates during an observation session).

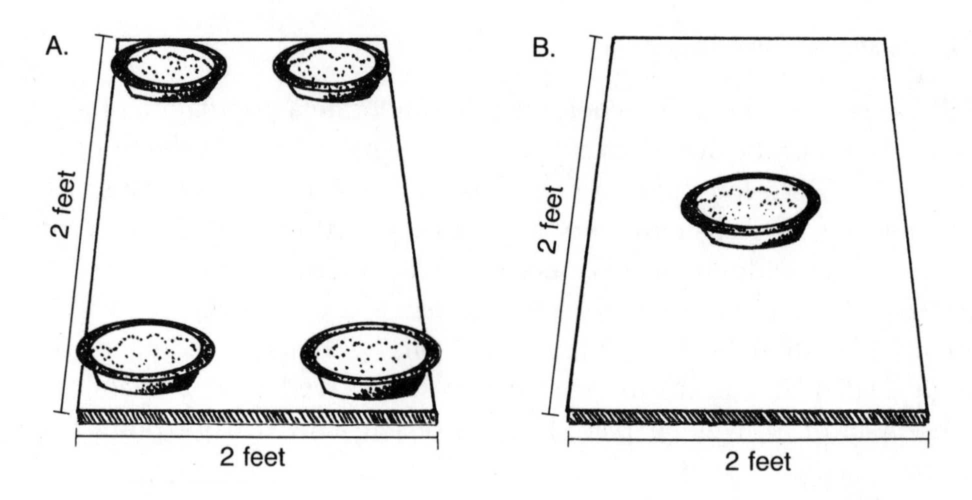

Figure 17.1. High food availability (A) and low food availability (B) platforms.

2. *Selecting an observation site.* Find a location in which gray squirrels are visibly feeding. Choose an observation site that is an open area with protective cover (for example, trees, bushes) bordering at least one side of the site. Observe the site for at least 1 hour to watch how the squirrels move about the area. This will help you to place the food platforms in the best location.
3. *Placing the feeding platforms.* Measure a distance 15 feet (4.5 m) from the protective cover. Place both the high-availability food platform (four pie plates) and the low-availability food platform (one pie plate) on the ground 15 feet (4.5 m) from the protective cover. Separate the two food platforms by 10 feet (3 m).
4. *Observing and recording feeding behavior.* Observe the squirrels as they feed over a five-day period, for approximately 2 hours per day. Record the number of animals feeding at any one time, the feeding rate, and the duration of feeding:

Number of animals feeding. At 5-minute intervals, scan the observation site and record the number of squirrels feeding at each platform.

Feeding rate. Observe one animal at a time, and record the number of seeds it eats in a 1-minute period. Make sure that you record feeding rates at both platforms.

Duration of feeding. Observe one animal at a time, and record the amount of time each squirrel spends eating (from the time it

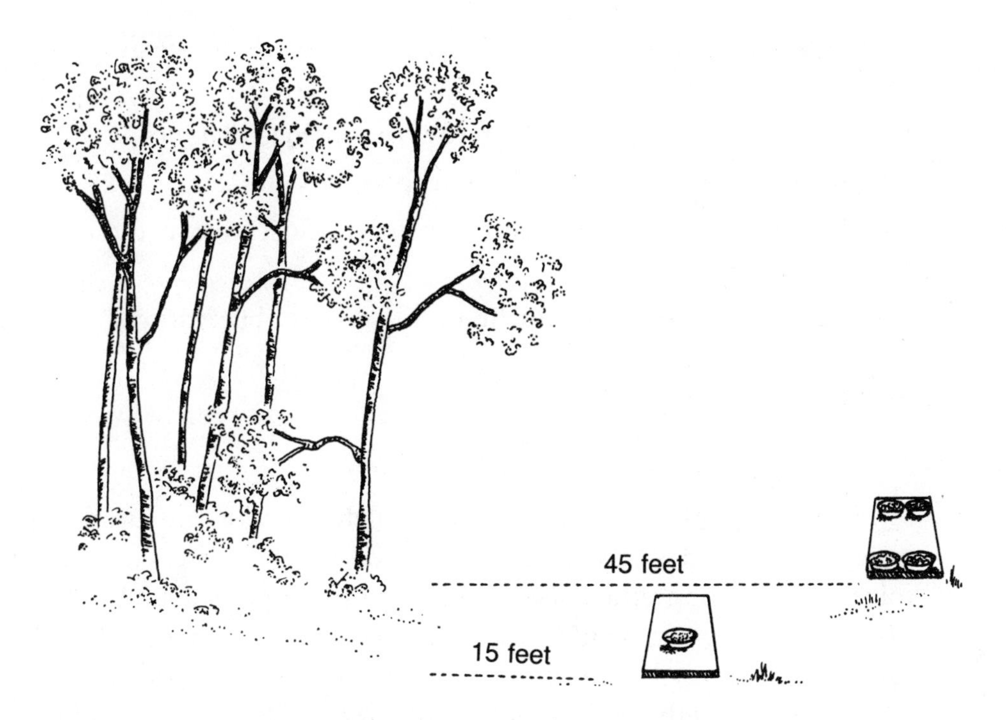

Figure 17.2. High food, far distance, with low food, close distance.

enters until the time it leaves the platform). Be sure that you record durations of feeding at both platforms.

5. *Moving the high-availability food platform.* Move the high-availability food platform to a spot 45 feet (14 m) from the protective cover. Leave the low-availability food platform in its original location (Figure 17.2). Repeat the observations in step 4 for five more days.
6. *Summarizing your data.*
   a. Calculate the average number of squirrels visiting each platform at a given time by dividing the total number of squirrels observed at each platform by the number of observations you made.
   b. Calculate the average feeding rate for each platform by summing the total number of seeds eaten by the number of minutes you recorded observations.
   c. Calculate the average duration of feeding at each platform by dividing the total time spent feeding by the number of squirrels you observed.

Summarize your data by making comparisons between the following conditions:

- high food, close distance, with low food, close distance.
- high food, far distance, with high food, close distance.
- high food, far distance, with low food, close distance (second observation).
- low food, close distance for first and second observations.

Questions

Feeding Strategy with a Low Threat of Predation

1. If the feeding platforms are located relatively close to protective cover (15 feet [4.5 m] away), one can predict that squirrels will maximize their feeding efficiency by going to the high-availability food platform. Thus, one would expect that the average number of squirrels, the feeding rate, and the duration of feeding would all be greater in the high food availability condition. With more areas to feed (four versus one), there should be less competition and greater opportunity to feed for longer durations. Are your results consistent with these predictions?
2. If your results were not consistent with your expectations, can you explain why you may have observed other behavior?

Feeding Strategy with a Greater Threat of Predation

1. When the high-availability food platform is moved 45 feet (14 m) away from protective cover, the squirrel's ability to run quickly for cover is greatly compromised. Will squirrels sacrifice feeding efficiency for greater safety? If one answers, yes, the following predictions can be made. The average number of squirrels, the feeding rate, and the duration of feeding should all decline for the high food availability condition when the feeding platform is moved farther from protective cover. Are your results consistent with these predictions?
2. When given a choice between high food availability at an unsafe distance and low food availability at a "safe" distance, one might predict that more squirrels would visit the single-dish platform. However, competition would be greater so the average feeding rate

and duration of feeding should decline. Are your results consistent with these predictions?

3. Overall, do your results suggest that the feeding strategy of gray squirrels involves a tradeoff between feeding efficiency and safety? How might seasonal pressures on squirrels' food supplies affect their behavior? Do you think that more squirrels are killed during periods of abundant food or scarce food?

CHAPTER 18

Hunting Behavior of Domestic Cats

The Ancient Egyptians treated cats as sacred animals, largely because of their superior hunting skills. Cats were kept in homes to guard against rodents and poisonous snakes. They were used to hunt small game and waterfowl. Cats had a status similar to other family members. They were mourned upon death, often mummified, and buried in special cat cemeteries that lined the banks of the Nile River. An excavation of one cemetery at the turn of the century uncovered over 300,000 mummified cats!

Although cats have been domesticated for thousands of years, they have retained their superior hunting and prey-capture skills. Their hunting instincts first appear at around six weeks of age. They start to practice their hunting abilities on their siblings, crouching, pouncing, batting, and biting each other.

Cats are solitary animals and prefer to hunt alone, usually at twilight or dawn. This is when potential prey (for example, mice) are most active and cats can make the best use of poor light conditions. At the back of their eyes, cats have a special layer of **iridescent cells,** which shift light and color. Under low-light conditions, these cells have a multiplying effect on the light received so that cats can see more with little light. The cat's iris contracts to a slitlike pupil under normal light conditions, shutting out excess light. Cats also have overlapping **visual fields,** which provide **depth perception,** giving them excellent accuracy in judging distances.

Whiskers also are important to a cat's ability to hunt its prey. The whiskers are highly responsive sense organs, moving forward when curious and backward when threatened or defensive. The whiskers are so sensitive that they can detect minor changes in air movements as the cat passes solid objects. It is thought that the whiskers are used to instantly check the body outline of prey objects and to direct the cat's bite to the victim's neck. Without them, a cat might misjudge its killing bite in the dark.

How does a cat hunt its prey? A cat immediately turns its head to the sight or sound of a prey object. Then, it flattens its body to the ground, glides nearer, and crouches down to watch the prey. Its extends its tail, the tip begins to twitch, and it turns its head from side to side. Then, suddenly, the cat releases built-up energy and pounces on the object, grasps it in its front paws, and bites it. This hunting sequence can be broken down into the behaviors that follow. In addition, two other behaviors associated with hunting, tooth rattling, and batting may be observed. These behaviors are described, as well.

Glide. The cat extends its head forward with its ears flattened (Figure 18.1A). Its hips, shoulder blades, and belly are kept low to the ground. It moves forward in long, fluid movements.

Figure 18.1. Glide, crouch, and tail-twitch postures of domestic cats.

Crouch. The cat pauses and hunches over with its tail out behind its body (Figure 18.1B).

Tail twitch. The tip of the cat's tail or its entire tail twitches rapidly from side to side (Figure 18.1C). The hips may swing back and forth as well.

Head turn. The cat stares at the prey object and then appears to sway its head from side to side.

Pounce. The cat springs into the air and grasps (or misses) the prey object with its claws fully extended.

Bite. The cat delivers a killing bite to the neck region.

Tooth rattle. The cat chatters its jaw while watching a prey object.

Bat. Before and/or after killing the prey, the cat may toss the prey, pat it, or bat it about.

Purpose

To observe and understand the hunting and prey-capture behavior of domestic cats.

Observational Setting

You will observe a cat in three different situations: tracking a moving object, observing a prey object, and attempting to catch a prey object. This project can be done in your home with your pet cat or in someone else's home if you do not have a cat. It is best if you can take your cat outdoors into a yard or a park for a portion of the observations. It is also best if your cat can observe birds in trees or other prey objects (for example, birds on a window bird feeder, pigeons on a window ledge) from a window of your house.

Materials

- ☐ an adult cat
- ☐ wind-up toy that would interest the cat
- ☐ notebook
- ☐ pencil
- ☐ stopwatch, or watch with second hand

Procedure

1. *Tracking a moving object.* Move about 6 feet (2 m) from the cat. Release the wind-up toy so that it moves across the floor. Note how long it takes the cat to approach the toy. (If the cat is scared of the toy or shows no interest in "hunting" it, try this portion of the experiment with a ball on a string.) Record all instances of the following behaviors you observe:

   Glide. How does it glide? Does it glide one or more times?

   Crouch. How does it crouch? How long does it crouch?

   Tail twitch. Does the tip twitch, or does the cat's whole tail and hips appear to twitch?

   Head turn. How many times does the cat turn its head? How long does it perform this behavior?

   Pounce. Estimate how far the cat leaps. Observe its hind legs, its front paws, and its claws.

   Bite. How many times does the cat bite? Does it always bite in the same place?

   Tooth rattle. How long does it perform this behavior?

   Bat. Does it bat the object before and/or after capturing it? How long does it perform this behavior?

2. *Observing a prey object.* Find a location in your house or apartment where the cat can sit on a table, chair, or window ledge and observe a prey object. Watching birds through a window at a bird feeder would be ideal (Figure 18.2). If you do not have a window feeder, you can try sprinkling seeds or bread crumbs on the window ledge. If necessary, you might need to make your observations outside. Locate a situation in which the cat can watch, but cannot possibly reach, a potential prey object in a tree, on a fence, and so on. Repeat the same observations listed in step 1.
3. *Observing prey outdoors.* Take the cat outdoors where you know there will be birds or small animals, such as chipmunks or squirrels, or large insects, such as grasshoppers. If you need to perform this portion of the experiment indoors, you can use a large grasshopper or cricket (which can be purchased at a pet store or bait store) as the

Figure 18.2. Cat observing birds at a window feeder.

prey object. Watch your cat as it locates potential prey objects. Repeat the same observations listed in step 1.

Questions

Locating a Prey Object

1. How did the cat notice the prey object in each of the three situations? Did it appear to use its vision, its hearing, or both?
2. How did the cat respond to the prey object when it was first detected? Did it just sit and watch? Did it move its ears in any way? Did it shift its posture?

Tracking and Hunting Behavior

1. Did you observe the same behaviors in all three situations? Were some behaviors observed more frequently? Were some longer in duration?

2. Did the cat appear to have a strategy in each of the three situations?
3. Was the cat successful when it attempted to catch some prey? If so, is the cat an experienced hunter?

Understanding the Functions of Behavior

1. Some scientists have suggested that the cat turns its head from side to side to better judge the distance of its prey. Would you agree with this hypothesis? Did the cat perform this behavior in all three situations?
2. How good do you think a cat's vision has to be to be a successful hunter?
3. Why do you think that a cat uses the gliding motion? Is there some other behavior that might be more effective?
4. Why might a cat crouch and wait before pouncing on its prey?
5. Scientists have suggested that the tooth-rattle behavior occurs in situations in which the cat can see its prey but not bite it. Did you see tooth rattling in any of the three situations? Would you agree with the scientists? If you did not observe this behavior, in which situation(s) would you have expected to observe it?
6. Tail twitching is a very common behavior that we have probably all observed in cats. Some suggest that it twitches because the cat cannot decide whether to stay still or rush forward. Others suggest that the cat is building up its energy to pounce. Which hypothesis do you agree with? What observations support your conclusion?
7. What can you conclude about hunting behavior in your cat? Would you expect to see the same behavior in all domestic cats? Why or why not?

CHAPTER 19

Behavior at the Bird Feeder

One animal that can be observed almost anywhere is the songbird. Whether you are in the city or the country, the sound of birds is never far away. Most people know how to recognize one or more kinds of birds, either by sight or sound. Songbirds are fascinating to watch as well. They exhibit a wide range of social behavior, communicated by sound, posture, and visual displays.

There is no better place to observe songbird behavior than at your own backyard bird feeder. A bird feeder provides a concentrated, safely located food source, unlike any other found in nature. As a result, normal social interactions are intensified as a variety of songbirds come to feed.

Depending on the species and the time of year, songbirds may visit a bird feeder alone, in male-female pairs, as a family, in a flock, or as a mixed flock of two or more species. Perhaps the best time to observe birds at your bird feeder is in winter. Survival is the most important thing to songbirds in the winter. Thus, you will find that many species will readily visit your feeder. In his book *The Habitat Guide to Birding,* McElroy lists songbirds that commonly visit winter feeders:

Common Winter Feeders	
Black-capped chickadee	Goldfinch
Bluejay	Hairy woodpecker
Brown-headed blackbird	House sparrow
Cardinal	Mockingbird
Carolina chickadee	Mourning dove
Common redpoll	Pine siskin
Downy woodpecker	Purple finch
Evening grosbeak	Red-bellied woodpecker
Fox sparrow	Red-breasted nuthatch

Red-winged blackbird
Slate-colored junco
Song sparrow
Starling
Tree sparrow
Tufted titmouse
White-breasted nuthatch
White-crowned sparrow
White-throated sparrow

Attracting Songbirds

To attract songbirds, you only need to make them feel welcome by providing a safe source of food. You can use many types of bird feeders and food sources. Different setups will attract different types of birds. For example, some birds, like snow buntings, prefer to feed in a cleared ground area. Others, like cardinals, will come to a shelf-feeder mounted outside your window.

You can purchase a commercial bird feeder or construct one of your own. A glass-topped shelf-feeder is ideal. Or you might hang a coconut with one end cut out to attract nuthatches and chickadees. A dead limb with holes filled with peanut butter can make an interesting feeder as well.

Make sure that you keep the food area clean and put out fresh food daily. In general, birds attracted to feeder food can be categorized as insect-eaters (for example, woodpeckers, warblers), seed-eaters (sparrows, juncos), or insect/seed-eaters (bluejays, chickadees). Insect-eaters prefer beef suet or peanut butter. Seed-eaters prefer sunflower seeds, hemp seeds, millet, and cracked corn.

Identifying Birds

It takes a while to figure out how to look at a bird in order to identify it. The best way to become skilled at identification is to write descriptive notes as you spot new birds. A bird's color, which seems to be the most obvious cue, tends to be the least reliable. Instead, look also at the comparative size of the bird, its shape, its head markings, the shape of its beak, its wings, its tail, its feet, and its breast markings. Try to observe its method of flight, and remember its call notes and song. With a good physical and behavioral description, you should be able to find the bird in an identification guide. Local libraries usually have a number of bird-watching guides that will assist you in identifying birds.

Observing and Understanding Bird Behavior

Birds are individualists. You will see birds that are docile, birds that are scrappy, belligerent birds, sociable birds, birds that are approachable, and birds that are aloof. They communicate their behavior predominantly through sound and vision.

The visual communication system of birds can be very subtle. An aggressive threat, for example, might be signaled by the mere fluffing of the head feathers. Thus, it is best to spend some time just observing birds, noting carefully what their normal feather and resting postures look like.

Probably the most common behavior you will observe at a bird feeder is aggression. When many kinds of birds are attracted to the same source, competition is inevitable. In his excellent book, *A Guide to Bird Behavior,* vol. I, Donald Stokes identifies four types of aggressive displays commonly seen at the bird feeder (Figure 19.1 illustrates three of

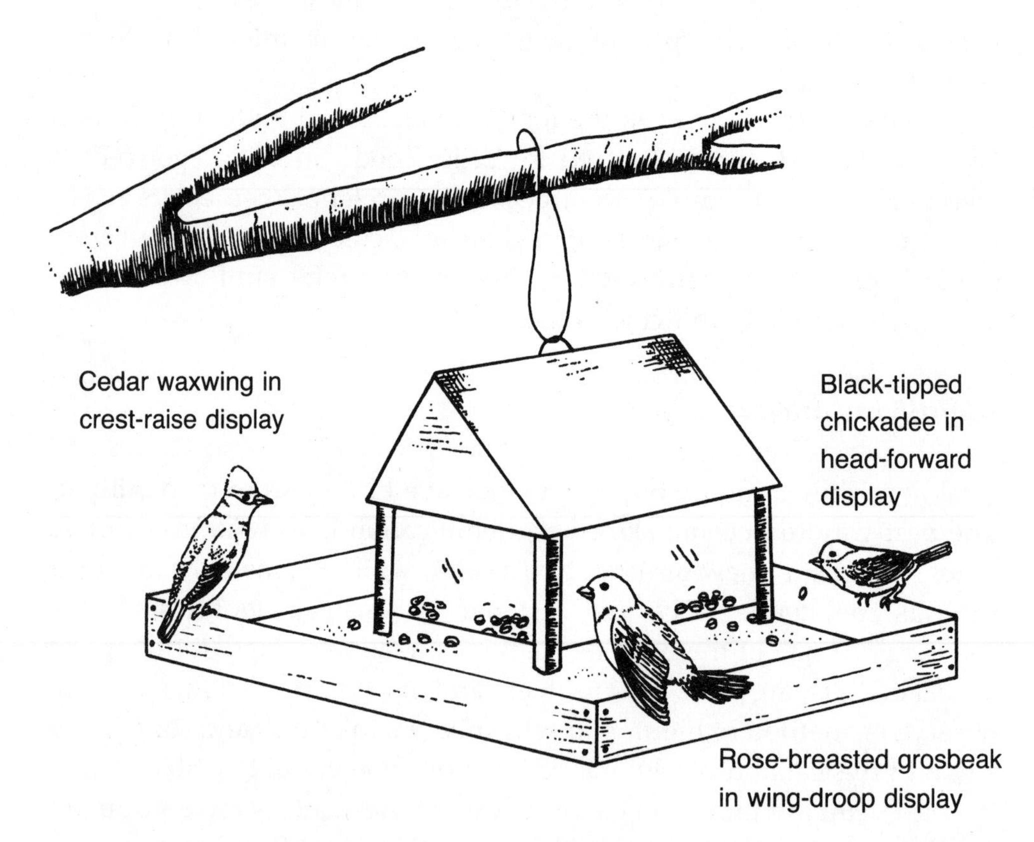

Figure 19.1. Crest-raise, wing-droop, and head-forward displays of birds.

these). Watching for these behaviors is an excellent starting point for your project.

Crest raise. Birds with a crest, like bluejays, raise their head feathers for 1 to 2 seconds when a bird lands near them. Birds without crests simply fluff their head feathers in a way that changes the profile of their head.

Wing droop. The aggressor bird raises its tail slightly. At the same time, its wingtips are lowered from the top of the base of the tail to a point below the tail. The more intense the display, the more the wings droop.

Head forward. The aggressor extends its body horizontally with its bill, either open or closed, extending straight forward.

Perch-taking. The aggressor flies at a perched bird, displacing it and taking its perch.

Many other behaviors may be displayed depending on the types of birds at the bird feeder. As you might expect, you will see differences in behavior among birds of the same species (for example, a male and female pair) and among birds of different species.

Purpose

To identify as many songbird species as possible and understand how they communicate through visual displays.

Observational Setting

For the purposes of this project, a shelf feeder mounted outdoors on the window ledge or hung close to a window has been used as a model; however, you may design and construct your own bird feeder.

Materials

- ☐ a shelf-type bird feeder
- ☐ 10 pounds (4.5 kg) of bird seed
- ☐ bird identification guide
- ☐ notebook
- ☐ pencil

- ☐ stopwatch, or watch with second hand
- ☐ binoculars (7.5 × 35 power), if available

Procedure

1. *Setting up a bird feeder.* Mount a shelf-type bird feeder on a window ledge, or hang it from a tree where you can easily observe birds coming and going (Figure 19.2). If the feeder is mounted on a window ledge, use a thin, sheer curtain to allow you to observe the birds without being easily detected. In either case, try to choose a location with nearby shrubbery or trees to create a sense of security for the birds.

   Stock your bird feeder with commercially available bird seed. Keep refilling the bird feeder when empty and leave it in place for several weeks before you begin your formal observations. This will permit the birds to become accustomed to the new feeder and a consistent food source.
2. *Identifying the birds.* Use a bird identification guide to familiarize yourself with the different types of birds that visit your feeder.
3. *Observing males and females.* Note differences, if any, between males

Figure 19.2. Shelf-type bird feeder.

and females of the same species. Try to determine whether each species typically feeds alone, in male–female pairs, or in flocks.

4. *Observing different groups.* Choose three groups of birds to watch:
   - male–female pairs.
   - two or more birds of the same species.
   - two or more birds of different species.

   Try to observe each kind of bird at rest. Note the natural resting positions of the wings, the head, the tail, and the bill. Then, begin to observe their interactions with other birds. Try to identify visual displays by looking for changes in the shape, movement, and posture of the wings, head, tail, and bill. Pay particular attention to what was happening before and after the behavior occurred. Record data on the following aspects of the behavior:
   - type of interaction observed (male–female, same species, or different species).
   - sex of birds (if known).
   - what was happening immediately before the interaction.
   - what was happening immediately after the interaction.
   - description of the behavior.
   - whether or not the behavior was reciprocated.
   - whether or not the behavior was repeated.

   Try to observe as many interactions as possible until you can identify six or more types of visual displays. Use a data sheet similar to the following one (a sample line of data is shown).

Date: ____________
Time: ____________
Weather Conditions: ____________
Food Source: ____________

Pair	Head	Bill	Wings	Tail	Reciprocate	Repeat	Notes
M–F	crest up	open	spread, flutter	up	Y	Y	appeared to be mated pair

Questions

Observing Visual Displays

1. Why do birds use visual displays rather than overt aggressive behavior?
2. Look over the data that you have gathered. Note what happened before a display occurred. What typically was the outcome of the threat display? Did the aggressor have to repeat a display more than one time?
3. Did you see a threat display reciprocated? You probably observed a number of threats but very little actual aggression. This is because most animals do not actually fight when competing over food. Actual aggressive encounters may draw attention to them, resulting in greater risk. By signaling its intentions with an aggressive threat display, a bird also minimizes its energy expenditure and maximizes its ability to feed.

Identifying Birds by Behavior

1. Did the behavior you observed help you better identify male and female birds? Sometimes when it is impossible to tell males from females on the basis of size, shape, and color, behavioral differences will help you distinguish males from females. You may have observed mate-feeding or courtship behaviors, involving males and females.
2. In species where you could distinguish males from females, were the males typically the aggressor or the dominant bird?
3. Did you see clear behavioral differences among birds of different species?
4. Were you able to identify different types of threat displays for different species?
5. Were you able to identify birds on the basis of their feeding behavior? Did some birds eat at the feeder while others took the food away to a tree to eat? Were there differences in the way different species ate seeds? For example, how did each of the species you observed crack sunflower seeds? All of these behaviors should give you further clues to the identity of the birds you observed.

CHAPTER 20

Bumblebee Searching Strategies

Few living things are as dependent on each other as bees and flowers. Flowers produce nectar and pollen, the two types of food needed by bees. Bees, in turn, fertilize flowers by transferring **pollen,** the flower's powderlike male sex cells, from one plant to another. A delicate balance exists between the two. Flowers need to produce **nectar,** a sticky, sugar-rich substance, to attract bees. Yet, if a bee could obtain a full meal on a single flower, **cross-pollination** would not occur.

Bumblebees have high energy requirements and typically visit twice as many blossoms per unit of time as other bees. Often, they visit more than 200 flowers in a single trip. Bumblebees collect nectar as their sole source of food energy. Pollen, which they frequently collect on separate trips, is rich in protein and is an essential food source for body growth.

Bumblebees need to be highly efficient in order to meet their food and energy requirements. What do they look for in flowers, and how do they search for **optimal food sources?**

Recent experiments with bees indicate that they learn the features of a flower more easily than they learn other things. They land instinctively on small, brightly colored objects. And they are particularly attracted to shapes that have a high ratio of edges to unbroken areas (for example, flowers shaped like daisies). The first thing they learn about a flower is how it smells. Next, they learn to distinguish color and then shape. Finally, they learn the time at which food is available and schedule their visits with "appointment book" accuracy.

Bees have color vision similar to that of humans, except their **observable spectrum** is shifted. Unlike humans, they cannot see very long wavelengths—they are **red blind.** However, bees can see very short wavelengths, including **ultraviolet light.** As a result, bees can distinguish three primary colors: yellow, blue, and ultraviolet. What a bee sees in a flower and what we see can be rather different. For example, to a bee, green leaves appear almost a colorless gray with a slight yellowish tint. This neutral background enhances the colorfulness of flowers for them.

Individual bumblebees also show clear color preferences. For example, there are two almost identical species of hawkweed that bloom at the same time. One is orange and the other is yellow. In field studies in which individual bees were tracked, some bees visited exclusively orange hawkweed, whereas others preferred the yellow flowers to the orange flowers with a ratio of over 10 to 1.

One of the most interesting features of a flower utilized by bumblebees is a **nectar guide** (Figure 20.1). This unique marking in the flower visually leads the bee to the source of nectar. Such a guide might be a radially arranged set of lines or a striking pattern of dots around the flower's "entrance." But nectar guides may not always be visible to the human eye. For example, the cinquefoil flower appears pure yellow to us. To bees, however, the outer portion of the petals reflects ultraviolet light. Thus, bees see a purple flower with a yellow nectar guide in the center.

How does a bumblebee approach a flower? A bee is first drawn from afar by a flower's odor and color. Then, it dips down on the edge of the flower, attracted by the line of contrast between the flower and the leaves. Following a nectar guide, the bee moves toward the center of the flower (Figure 20.2) and probes with its **proboscis,** a tubular organ used for sucking nectar. If successful, the bee moves quickly from blossom to blossom, sucking nectar into its **honey sack.** After going back to the nest with a full honey sack, the bee often returns to the same patch of flowers.

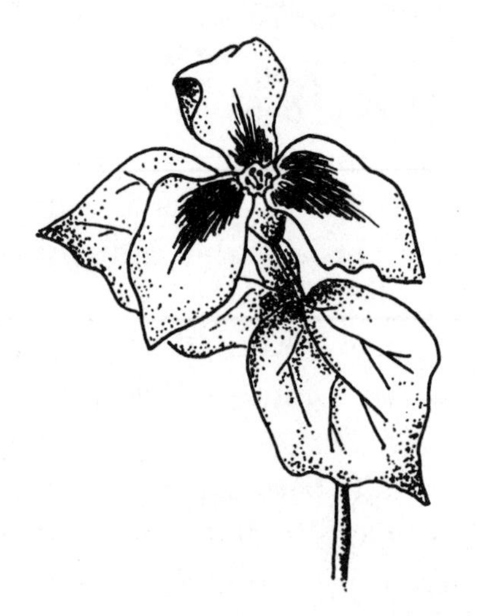

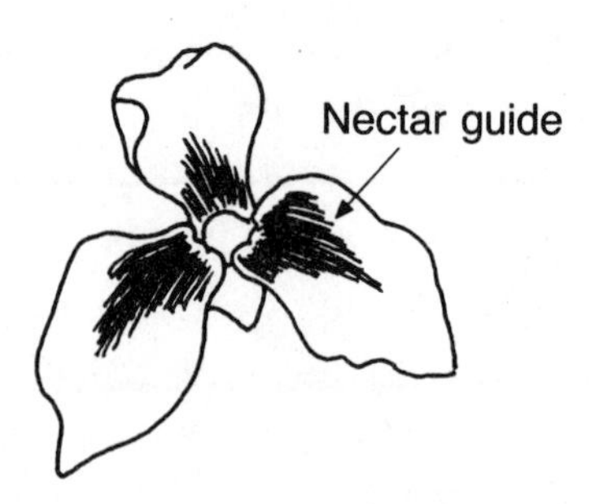

Figure 20.1. Nectar guide.

Figure 20.2. Bumblebee approaching from the edge of a flower.

Purpose

To observe some of the strategies bumblebees use in searching for and collecting nectar.

Observational Setting

Bumblebees feeding on nectar are best observed from late spring through summer. They are large bees and are easy to spot. An ideal observational setting is a field of wildflowers in which a number of different species are blooming at the same time. You also can observe them feeding in a planted flower bed or in any other patch of different kinds of flowers. If you live in a large city, a botanical garden is an excellent place to watch bumblebees.

Materials

- ☐ notebook
- ☐ pencil
- ☐ field guide to wildflowers (optional)
- ☐ stopwatch, or watch with second hand

Procedure

1. *Locating flowers and bumblebees.* Locate a field of wildflowers or another observation site with two or more flower species. Make sure you can spot some bumblebees.

2. *Observing flowers.* Spend some time observing the different types of flowers. Construct a data sheet listing the different types of flowers across the top, like that shown in step 4. If you have access to a field guide to wildflowers, try to identify each flower. If not, just qualitatively describe each flower for future reference. Include a description of the following features:
   - color of the flower.
   - shape of the flower.
   - description of any visible nectar guides.
   - single or multiple blossoms per plant.
3. *Observing bumblebees.* Watch until you observe a bumblebee fly into the area and alight on a blossom. Record the time. Follow the bumblebee as closely as possible, counting every time it collects nectar from another blossom. Note whether it lands on the edge or center of the blossom (this may be difficult if it is a small flower). When the bee flies away from the area, record the time again (see sample data sheet). This will give you an approximate duration for the collection trip. Repeat this procedure until you have observed at least ten bumblebees collecting nectar.
4. *Observing bumblebees with different flowers.* Wait approximately two

Sample Data Sheet
Bumblebee #1

Date: July 28
Begin collection trip: 3:05 P.M.
End collection trip: ____________

Yellow Hawkweed	Orange Hawkweed	Cow Vetch	Evening Primrose
EEEEE EEEEE	EC	C	

Notes: An E is used to indicate that the bee landed on the *edge* of the flower; a C is used to indicate that the bee landed on the *center* of the flower. It is easier to tally your data if you record instances in blocks of five, as was done in the column for yellow hawkweed.

weeks until different flower species are in bloom. Repeat your observations on ten more bumblebees.

Questions

Bumblebee Strategies

1. Review the data you collected. Calculate the percentage of time each bee collected nectar from the different flower types on a single trip. Even though there may have been many species of flowers, did the bees tend to collect nectar from the same type of flower? Did all the bumblebees you observed in a single afternoon collect nectar from the same kind of flower? What features of the flowers might help you explain your results?
2. Did you ever observe the same bumblebee return to a blossom from which it already collected nectar? Was this a very common occurrence? What cues do you think a bumblebee uses to prevent "searching" a blossom repeatedly?
3. Did individual bumblebees visit each blossom on a plant before moving to the next plant? Or did they move back and forth between plants?
4. Were other bumblebees collecting nectar nearby?
5. Approximately how far did the bumblebee travel while collecting nectar in a single trip?
6. Did you see evidence that bumblebees were using a strategy to forage efficiently with a minimum amount of effort? Give several examples of how you think the bees were conserving energy.
7. Did you see any changes in strategy when different kinds of flowers were in bloom?

Features of the Flowers

1. What color were the flowers the bumblebees visited? Did they always visit the same color flower?
2. What was the shape of the flowers they visited?
3. Did the bees tend to alight on the edge or the center of the blossoms? Did their landing strategy appear to be related to the shape of the flower? The size of the flower? Or the presence of visible nectar guides?
4. Do you think the physical features of the flowers affected the search and collection strategies you observed?

Appendix: Animal Cross-Reference

If you have found a project of interest but cannot locate the appropriate animal(s) to observe, use this animal cross-reference to locate a similar animal to observe. For example, suppose you are interested in the feeding strategies of gray squirrels, but none are nearby your home. In the left column of the table, find Chapter 17, Gray Squirrels. There you will find that you can perform a similar project with red squirrels or fox squirrels.

Chapter	Animal	Substitute Animals and Notes
1	Fireflies	None, but you can find fireflies in almost any habitat.
2	Primates—chimpanzees, gorillas, orangutans	Some New World monkeys (e.g., *Rhesus* monkeys).
3	Wolves	Coyotes, foxes, dingos, and dogs.
4	Siamese fighting fish	African cichlid fish; aggressive displays can be observed in peacocks, turkeys, and ruffed grouse.
5	Crickets	Grasshoppers, but their song will sound more mechanical.
6	Horses	Any hooved animal that is a "follower," including sheep, goats, cows, and llamas. It might also be interesting to try project with "hiders," such as deer and giraffes.
7	Mallard ducks	Black duck is the most similar, but you also could observe gadwalls, pintails, widgeons, and teals.
8	Human infants	None, but you could try a similar experiment using great apes such as chimpanzees, gorillas, or orangutans.

9	Lambs	Any hooved animal that is a "follower," including sheep, goats, cows, and llamas.
10	Bluejays	Any of the larger species of songbirds.
11	Neon tetra fish	Any species of tropical aquaria fish that school such as zebra fish. You also could try to do an observational study at an aquarium. Ask the curator for the best species or tanks to observe.
12	Chickens	Ducks, pigeons, and shorebirds.
13	Domestic dogs	None, but you might attempt the same study with young puppies and compare your results to those with adult dogs. You also could attempt to look for parallel types of behaviors in other domestic pets.
14	Pigeons	Mourning doves and shorebirds.
15	Human adults	None, but you could do a similar study with verbal toddlers or preschoolers.
16	Garden spiders	Other orb-weaving spiders and funnel web and sheet web spiders.
17	Gray squirrels	Red squirrels and fox squirrels; you also could observe ground squirrels, but their protective cover would have to be on the ground or underground.
18	Domestic cats	None, but you might attempt the same study with young kittens and compare your results to those with domestic cats.
19	Songbirds	Pigeons.
20	Bumblebees	Honeybees and possibly butterflies.

Glossary

aggressive display Visual and/or auditory behaviors that signal the intention to aggress or attack an opponent.

aggressor An animal that initiates a conflict with another individual.

alpha animal The top-ranking animal in a social structure. All other animals defer to that animal. *See also* **dominant.**

altricial Newborn or newly hatched animals that are completely dependent on their mother for care.

amniotic fluid Watery liquid contained in the innermost sac enclosing the embryo of a mammal, reptile, or bird.

anthropomorphism Giving humanlike qualities or characteristics to nonhuman animals or inanimate objects.

auditory signals Communications used between animals based on sound.

beta animal The second-ranking animal in a social structure. Only the alpha animal is more dominant. *See also* **dominant animal.**

brood To sit on and hatch eggs, or to cover or hover over young. Also used to refer to a group of young all hatched at the same time.

capture threads *See* **spiral coils.**

carnivore An animal that eats flesh.

chemical signals Communications between animals based on smell.

clutch A nest of eggs or a brood of young chicks. *See also* **brood.**

comb A red, fleshy growth on the top of a chicken's or rooster's head.

controls Aspects of a scientific experiment that are held constant, or are not varied.

courtship displays Visual and/or auditory behaviors that signal sexual interest or receptivity to a potential mate. *See also* **receptive posture.**

cross-pollination The transfer of pollen from one flower to another.

data Records or notes on the observations or measurements made in a scientific study.

data sheet A specially designed form for recording scientific data.

debrief To explain the purpose of a scientific study to humans following their participation in a study.

depth perception The ability to see an object in perspective or at varying distances.

dominance hierarchy A ranking of individuals based on their dominance status.

dominant animal The highest ranking animal in a social structure or behavioral interaction. Usually the dominant animal has the greatest access to food, potential mates, nesting sites, and so forth.

drag line A single strand of silk by which the spider attaches itself to an orb web.

electrical signals Communications between animals based on electric currents; typically communicated between fish in an underwater environment.

ethologist One who studies behavior patterns of animals.

feeding efficiency A strategy used by animals that minimizes the amount of energy expended in obtaining the maximum amount of food.

field study A study of behavior conducted in an animal's natural environment or a re-creation of their natural environment.

file A hardened vein under a cricket's wing used in song production.

fledge To learn to fly.

flock A group of animals that live, feed, and move together.

follower strategy Strategy used by hooved mammals whereby young stay close to their mothers at all times. Opposite of "lying out" strategy.

funnel web A platformlike spider web with a tube or funnel used by the spider to capture prey and exit.

habitat Characteristics of the environment where an animal or a plant naturally lives or grows.

hackles The neck feathers of a chicken, which it raises to warn another chicken not to come closer. Hairs on a dog's neck and back that are erected when the dog threatens to fight.

herbivore A plant-eating animal.

home range The physical area in which an animal is born, lives, eats, sleeps, and dies.

honey sack An area into which bee sucks nectar while collecting it.

hub Meshlike center of an orb web. *See also* **orb web.**

hypothesis An educated guess or prediction as to the outcome of a scientific experiment.

imprinting Inborn learning mechanism by which young learn, for example, to attach themselves to moving objects in the first few hours of life. *See also* **inborn.**

inborn Inherited characteristic or characteristic of an animal present at birth.

incubation Period during which an animal sits on its eggs until they hatch.

interaction markers Nonverbal body movements used to signal turn-taking in human conversations.

iridescent cells Light-producing cells found in cat eyes.

irregular mesh web A spider web of no particular design—usually just a mass of silk web.

journal article Written description of a scientific experiment. Any scientist should be able to read the article and replicate the experiment.

linear dominance hierarchy A ranking of individuals in which animal A is dominant over animal B, who is dominant over animal C, and so on.

loafing area An area of ducks' territory used for resting and preening.

lying out strategy Strategy used by hooved mammals whereby their young are hidden under protective cover for the first two weeks of their lives. Opposite of "follower" strategy.

maternal behavior Behavior exhibited by a mother toward her offspring.

molt To shed the outer layer of skin, hair, horns, or feathers.

mounting Process by which a male animal climbs on a female for the purpose of copulation.

naturalistic observation An observation done in the field in which no variables are manipulated or controlled. *See also* **field study.**

nectar Sweet liquid found in the center of a flower.

nectar guide Unique marking on a flower that guides insects to the source of nectar. It may or may not be visible to the human eye.

nestling phase The period during the development of young birds in which they are cared for by their mother in the nest.

nestlings Newly hatched birds that are confined to a nest.

nonlinear dominance hierarchy A ranking of individuals in which animal A may be dominant over animal B, and animal B may be dominant over animal C, but animal C may be dominant over animal A.

nonverbal communication Use of the human body, demeanor, and facial expressions to convey information.

observable spectrum Colors that may be observed by an animal.

olfactory sense An individual's sense of smell.

omega animal The lowest ranking animal in a social structure.

optimal food source A food source of high nutritional value that can be obtained with minimal risk of predation. *See also* **predation.**

orb web A traditional spider web, organized with radii and spiral coils. *See also* **radii** and **spiral coils.**

ovipositor A long tube extending from the rear of a female cricket's body.

pecking order Another term for a dominance hierarchy in chickens or other birds that is based on who pecks whom.

pollen Powderlike male sex cells found in the center of a flower.

precocial Newborn or newly hatched animals that are fully developed or nearly self-sufficient at birth.

predation Process of feeding on other animals (prey) in order to survive.

predator An animal that captures and feeds on other animals.

proboscis A tubular organ of bees for sucking nectar.

psychologist One who studies behavior and mental processes.

radial lines (radii) Lines of spider silk anchoring an orb web; they converge at the center or hub of the web.

receptive posture Position assumed by a female animal to signal her readiness for mating, nursing, and so forth.

red blind Inability of an individual to discriminate the color red.

replicable The ability to repeat exactly the results of a scientific experiment.

ritualized fighting Assumption of fighting postures and sparring of two individuals without actual fighting; typically follows aggressive displays.

sample A piece or portion of a group that reflects the characteristics of the whole group.

school A group of fish that swim together.

scientific method A method used to systematically collect and study scientific data.

scraper A rough surface on a cricket's wing used in song production.

sheet web A platformlike spider web under which the spider clings, waiting for prey to fall.

social status An individual's rank or position within a social grouping.

social structure Any form of organization of individuals, such as a flock, a school, or a pack.

spatial relationship The distance maintained between two individuals.

species Biological classification for different kinds of animals.

spiral coil Sticky spider silk that is attached to the radii of an orb web and is used for trapping insects.

subordinate animal An animal that defers to a dominant animal. The subordinate animal may show submissive behavior, characterized by body posture or coloration changes, as well as avoidance behavior.

syllable A sound produced by crickets when their wings are closed and rubbed together. A typical chirp is a train of three or four syllables.

systematic study Planned or orderly observation using a particular method.

tensile strength Resistance to stress.

territorial boundaries The physical boundaries of an area defended by an animal.

territoriality The tendency of an animal to defend an area such as a nesting or feeding site.

time-sampling method Naturalistic observation method by which one obtains a representative sample of behavior by observing at fixed intervals of time.

ultrasonic calls Communications between animals that are at frequencies above those detectable by the human ear.

ultraviolet light Light that is just beyond the violet end of the visible spectrum.

variables Aspects of a scientific study that are manipulated by the experimenter.

visual field The area that can be seen by an individual's eye.

wildlife biologist One who studies the origin, physical characteristics, habits, and so on, of wild animals.

wingpads Pre-wing area of immature crickets.

x-axis The horizontal plane of a two-dimensional graph.

y-axis The vertical plane of a two-dimensional graph.

zoologist A biologist who studies the life, classification, structure, and so on, of animals.

Selected References

Chapter 1

Lloyd, J. E. 1980. Photuris fireflies mimic sexual signals of their females' prey. *Science* 210:669–71.

Stokes, D. 1983. *A guide to observing insect lives.* Boston: Little, Brown and Company.

Chapter 2

Alcock, J. 1989. *Animal behavior.* 4th ed. Sunderland, Mass.: Sinauer Assoc.

Chevalier-Skolnikoff, S. 1973. Facial expression of emotion in non-human primates. In *Darwin and facial expression,* ed. P. Ekman. New York: Academic Press.

Morris, D. 1977. *Manwatching.* New York: Harry N. Abrams.

Redican, W. K. 1982. An evolutionary perspective on human facial displays. In *Emotion in the human face,* ed. P. Ekman. London: Cambridge University Press.

Chapter 3

Alcock, J. 1989. *Animal behavior.* 4th ed. Sunderland, Mass.: Sinauer Assoc.

Mech, L. D. 1970. *The wolf.* Minneapolis: University of Minnesota Press.

Sutton, A. and M. Sutton. 1992. *Eastern forests.* New York: Alfred A. Knopf.

Chapter 4

Herald, E. S. 1972. *Living fishes of the world.* New York: Doubleday.

Hinde, R. A. 1966. *Animal behavior: A synthesis of ethology and comparative psychology.* New York: McGraw-Hill.

Morris, D. 1990. *Animal watching.* New York: Crown Publishers.

Ommanney, F. D. 1963. *The fishes.* New York: Time.

Chapter 5

Huber, F., and J. Thorson. 1985. Cricket auditory communication. *Scientific American* 253:60–68.

Matthews, R. W., and J. R. Matthews. 1978. *Insect behavior.* New York: John Wiley and Sons.
Simon, S. 1979. *Pets in a jar.* New York: Puffin Books.
Thornhill, R., and J. Alcock. 1983. *The evolution of insect mating systems.* Cambridge, Mass.: Harvard University Press.

Chapter 6

Crowell-Davis, S. L. 1986. Spatial relations between mares and foals of the Welsh pony (*Equus caballus*). *Animal Behaviour* 34:1007–15.
Morris, D. 1988. *Horsewatching.* New York: Crown Publishers.
Simpson, G. G. 1951. *Horses.* New York: Oxford University Press.
Walther, F. R. 1984. *Communication and expression in hoofed mammals.* Bloomington, Ind.: Indiana University Press.

Chapter 7

Barnett, S. A. 1981. *Modern ethology.* New York: Oxford University Press.
Holderread, D. 1978. *Raising the home duck flock.* Pownal, Vt.: Garden Way Publishing.
Sluckin, W. 1973. *Imprinting and early learning.* Chicago: Aldine Publishing Company.
Stokes, D. W. 1979. *A guide to bird behavior.* Vol. I. Boston: Little, Brown and Company.

Chapter 8

Bornstein, M. H., and M. E. Lamb. 1988. *Developmental psychology: An advanced textbook.* 2nd ed. Englewood Cliffs, N.J.: Lawrence Erlbaum.
Bower, T. G. R. 1989. *The rational infant.* New York: W. H. Freeman.
Bremner, J. G. 1988. *Infancy.* New York: Basil Blackwell.
Corter, C. M., and A. S. Fleming. 1990. Maternal responsiveness in humans. *Advances in the Study of Behavior* 19:83–136.

Chapter 9

Bradbury, M. 1977. *The shepherd's guidebook.* Emmaus, Pa.: Rodale Press.
Vince, M. A. 1993. Newborn lambs and their dams: The interaction that leads to suckling. *Advances in the Study of Behavior* 22:239–68.
Walther, F. R. 1984. *Communication and expression in hoofed mammals.* Bloomington, Ind.: Indiana University Press.

Chapter 10

Burton, R. 1985. *Bird behavior.* London: Granada Publishing.
Headstrom, R. 1970. *A complete field guide to nests in the United States.* New York: Ives Washburn.

Skutch, A. F. 1976. *Parent birds and their young*. Austin, Tex.: University of Texas Press.

Stokes, D. W. 1979. *A guide to bird behavior.* Vol. I. Boston: Little, Brown and Company.

Chapter 11

Ommanney, F. D. 1963. *The fishes*. New York: Time.

Chapter 12

Limburg, P. A. 1975. *Chickens, chickens, chickens.* Nashville, Tenn.: Thomas Nelson.

Chapter 13

Monks of New Skete. 1991. *The art of raising a puppy.* Boston: Little, Brown and Company.

Morris, D. 1987. *Dogwatching*. New York: Crown Publishers.

Stokes, D., and L. Stokes. 1986. *Animal tracking and behavior.* Boston: Little, Brown and Company.

Chapter 14

Fabricius, E., and A.-M. Jansson. 1963. Laboratory observations on the reproductive behaviour of the pigeon (*Columbia livia*) during the pre-incubation phase of the breeding cycle. *Animal Behaviour* 11:534–47.

Stokes, D. W. 1979. *A guide to bird behavior.* Vol. I. Boston: Little, Brown and Company.

Chapter 15

Argyle, M. 1972. Non-verbal communication in human social interaction. In *Non-verbal communication,* ed. R. A. Hinde. Cambridge: Cambridge University Press.

Hinde, R. A., ed. 1972. *Non-verbal communication.* Cambridge: Cambridge University Press.

Knapp, M. L. 1972. *Non-verbal communication in human interaction.* New York: Holt, Rinehart & Winston.

Chapter 16

Gertsch, W. J. 1979. *American spiders.* 2nd ed. New York: Van Nostrand Reinhold.

Jones, D. 1983. *The Larousse guide to spiders.* New York: Larousse and Company.

Kaston, B. J. 1972. *How to know the spiders.* 2nd ed. Dubuque, Iowa: William C. Brown.

Vollrath, F. 1992. Spider webs and silks. *Scientific American* 266:70–76.

Chapter 17

Newman, J. A., and T. Caraco. 1987. Foraging, predation hazard and patch use in grey squirrels. *Animal Behaviour* 35:1804–13.

Stokes, D., and L. Stokes. 1986. *Animal tracking and behavior.* Boston: Little, Brown and Company.

Van Wormer, J. 1978. *Squirrels.* New York: E. P. Dutton.

Wauters, L., and A. A. Dhondt. 1992. Spacing behavior of red squirrels, *Sciurus vulgaris*: Variation between habitats and the sexes. *Animal Behaviour* 43:297–311.

Chapter 18

Morris, D. 1987. *Catwatching.* New York: Crown Publishers.

Sayer, A. 1979. *The encyclopedia of the cat.* New York: Crown Books.

Stokes, D., and L. Stokes. 1986. *Animal tracking and behavior.* Boston: Little, Brown and Company.

Chapter 19

Booth, E. S. 1962. *Birds of the east.* Escondido, Calif.: Outdoor Pictures.

McElroy, T. P., Jr. 1974. *The habitat guide to birding.* New York: Alfred A. Knopf.

Socha, L. O. 1987. *A bird watcher's handbook.* New York: Teale Books.

Stokes, D., and L. Stokes. 1987. *The bird feeder book.* Boston: Little, Brown and Company.

Chapter 20

Heinrich, B. 1973. The energetics of the bumblebee. *Scientific American* 228:97–101.

Peterson, R. T., and M. McKenny. 1968. *Wildflowers.* Boston: Houghton Mifflin Company.

Stokes, D. 1983. *A guide to observing insect lives.* Boston: Little, Brown and Company.

Tinbergen, N. 1960. *Curious naturalists.* New York: Basic Books.

von Frisch, K. 1971. *Bees.* Ithaca, N.Y.: Cornell University Press.

von Frisch, K. 1953. *The dancing bees.* New York: Harcourt, Brace & World.

Index

A

B

C

D

E

F

G

H

I

J

L

Q

R

S

T

U

V

W

X

Y

Z